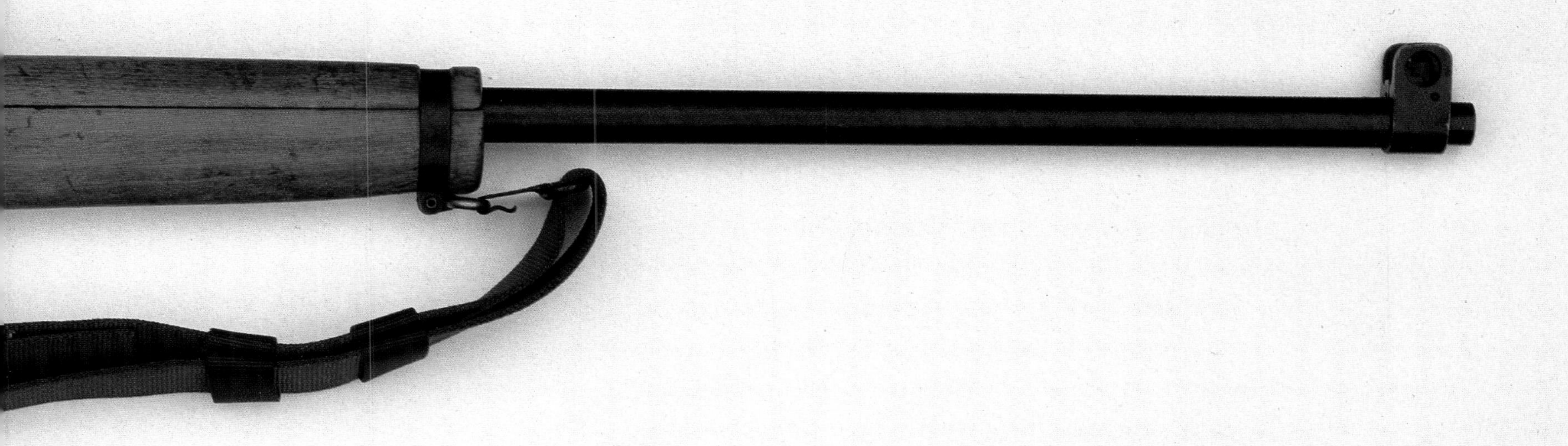

The New Encyclopedia of HANDGUNS

This book was devised and produced by
Multimedia Publications (UK) Ltd.

Editors: Valerie Passmore, Andie Oppenheimer
Production: Karen Bromley
Design: John Strange Associates
Picture Research: Military Archive and Research Services

First published in the United States of America 1986 by Gallery Books, an imprint of W. H. Smith Publishers Inc., 112 Madison Avenue, New York, NY 10016

ISBN 0 8317 6323 X

Typeset by Computerised Typesetting Services Limited, London
Origination by David Bruce Graphics Limited
Printed by Cayfosa, Barcelona, Spain

The New Encyclopedia of HANDGUNS

Christopher Chant

GALLERY BOOKS
An Imprint of W. H. Smith Publishers Inc.
112 Madison Avenue
New York City 10016

Half title
Typical of the high standards of presentation firearms is this beautiful silver-handled Smith & Wesson revolver, upgraded by Holland and Holland in a special case complete with tools and accessories

Title page
The revolving pistol came of age with the Colt types, at last giving individuals multiple-shot capability and ease of reloading

Above
Broken-frame revolvers were immensely strong and reliable and thus well suited to front-line military service

Contents

A sketch by an unknown artist shows men of the Austrian imperial musketeers. Their importance on the battlefield increased because wealthier employers were now making efforts to standardize weapons even in the days before mass production.

Small arms up to the Napoleonic Wars

Introduction

Small arms can best be defined as projectile-throwing weapons for one or occasionally two men using some type of controlled detonation to produce the gases that accelerate a small projectile through and out of the barrel. Other aspects of any such individual weapon are its furniture (for handling the weapon properly), a trigger mechanism, a method of reloading and a sighting system. So much is comparatively straightforward but, as will become clear, the ramifications and diversifications possible within this simple exposition were, are and will continue to be enormous as designers and operators alike seek to produce weapons with significant advantages over their likely opposition.

Below:
The lock and trigger of a typical weapon of the flintlock type, signed 'Bernhard Ortner' by the maker and showing the nature of the engraving (on the jaws of the lock) much favored for better pieces.

Early days

The first propellant used in firearms was gunpowder. Despite its great historical and military significance, its origins remain largely obscure other than the fact that the Chinese were apparently using it by AD 1000. The first extant references to gunpowder in Western literature occur in the middle of the 13th century, in two works by the celebrated British natural philosopher Roger Bacon and one by the German philosopher Albert Magnus.

The origins of firearms are also wrapped in mystery. It is clear that the Chinese and Arabs knew at an early date how to make and use 'roman candles' (probably based on bamboo tubes) loaded with layers of gunpowder and incendiary material to start fires at a distance for military purposes.

Above:
A German horseman uses a "hand cannon", though the extent of the muzzle flash seems slightly exaggerated! The aiming of such a weapon from the back of a horse was extremely difficult, even at close ranges.

Left:
The Abyssinian Emperor Theodore takes his own life at the end of the disastrous Battle of Magdala in 1868: a stylized Victorian depiction featuring the type of obsolescent flintlock still used by many "backward" countries at a time of great technological progress.

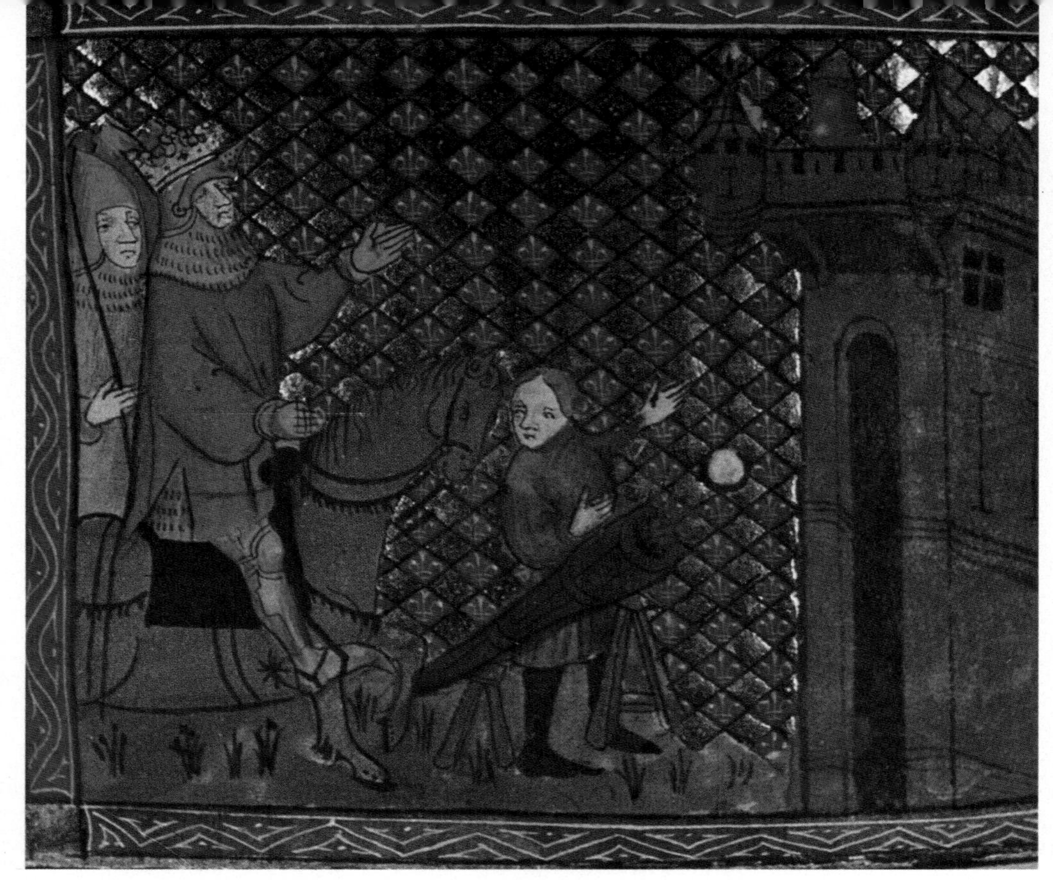

Above:
A late 14th-century French manuscript depicts the siege of La Rochelle by Louis VIII in 1224. Inclusion of a cannon is sheer fancy as such weapons did not exist at the time—the type of weapon shown is more typical of mid-15th-century weapons.

But details of this weapon are lacking, as too are any clear indications of the first use of genuine firearms for propelling projectiles from a tube. There are claims that the Moors used such a cannon in 1247 for the defence of Seville, and that the German town of Amberg possessed a rudimentary cannon in 1301, but these claims (especially that for the Moors) seem overly ambitious. Thus the first documented example of a gunpowder weapon is the cannon illustrated in the Millimete manuscript of 1326: the barrel is pot-shaped and mounted on a four-legged table, and the projectile is large and arrow-shaped, with base-mounted fins. There are other indications that such pot cannon were used at Ghent in 1313 and at Metz in 1324. From this it would seem that cannon were moderately widely used by the first quarter of the 14th century, indicating that gun protagonists must have overcome the enormous technical problems associated with gun founding and with the manufacture of gunpowder during the second half of the 13th century.

A similar state of uncertainty reigns about handguns, though there is good evidence that such weapons had been developed by the first quarter of the 14th century in Germany and in Italy. What is indisputable, however, is that all the gunpowder weapons of the period were of very limited capabilities: they were poorly made as a result of the limitations in the foundry techniques of the period, were very heavy, and had only small ranges because of the difficulty of matching the ball (often made of stone) to the bore of the weapon. Yet such weapons were effective, generally as a result of the shock effect of their noise and considerable impact at close ranges, and spurred their designers to greater efforts in terms of reliability, range and projectile velocity.

Below:
A 15th-century manuscript reveals a typical matchlock fowling piece of the day. Note the man's powder horn on the hip, and the match held in his right hand above the touch-hole.

Left:
A fanciful illustration from a manuscript of Marco Polo's journey to China purports to show the Khan's soldiers attacking the stubborn city of Sa-yan-fui in the province of Manyi; it actually shows a scene typical of European sieges of the period, complete with a trestle-mounted light cannon in support of crossbowmen and pikemen.

Below:
The idealized Death of Colonel Moorhouse, Madras Artillery, at the Siege of Bangalore *presents an odd view of how contemporary muskets could have been used. The weapon's short range is emphasized, albeit inadvertently.*

Above:
A German 16th-century military illustration depicts the pros and cons of matchlock and lance. The former had considerable effect on even an armored cavalryman, but was slow to fire and prone to misfire, while the latter had good reach at close quarters.

Left:
The petronel cannon-lock was a mighty weapon in its time, but weighty and difficult to aim. This illustration from about 1450 shows how a mounted soldier could cope with the problems of so unwieldy a weapon, using a saddle-braced rest and a cord support round the neck—leaving one hand free to aim the weapon and the other hand to hold the vital match.

Right:
A painting of the uniforms of three Guards regiments in 1751 shows clothing details in immaculate form, but the painter was less familiar with the details of the weapons, and has fudged the issue: the man on the left has his hand over the lock, the man in the center keeps the lock of his weapon in shade, and the man on the right has turned his flintlock musket round.

The cannon lock

The earliest weapons were of the 'cannon lock' variety, in which the muzzle-loaded weapon was touched off by the ignition (by a hot iron or lighted coal) of a small powder priming charge in the touch-hole leading down to the main propellant charge in the rear of the barrel behind the wadding and the ball. The barrel itself was normally made of cast bronze or brass (though wrought iron was also used), and gunpowder was not granulated and had a low saltpetre content. Combined with a badly fitting ball or bolt and relatively ineffective wadding, a moderately slow burn and pressure build-up resulted, rather than an instantaneous detonation which would have completely undone the laborious work of the gun founder. Such cannon-lock firing was used in both artillery weapons and in the first handguns, which appear to have been small artillery pieces with the barrel attached to the end of a pole supported at the rear under the firer's right arm and at the front by the left hand; the powder in the touch-hole was ignited with the right hand. The similarity between artillery pieces and handguns indicates that the two types were in all probability designed and used in parallel.

It seems that the cannon lock was used for at least 50 years, and though improvements were made in powder and casting techniques so that more effective barrels were produced, the basic design of the pole weapons did not develop significantly.

The matchlock

The next important step in the evolution of handguns was the German invention of the matchlock in the late 14th or early 15th century. In such a lock a length of burning match (hemp cord soaked in a mixture of saltpetre and other chemicals so that it burned steadily yet slowly) was held in the upper end of a C-shaped bar hinged to the barrel at its lower end so that the firer could simply depress the bar with a finger to bring the burning match into contact with the priming charge in the touch-hole. This enabled the handgun to be held with both hands, making possible more accurate aiming and opening the way for the development of sights. From this stemmed the development of weapons with shaped butts, and in the course of the next 50 or so years the matchlock transformed handguns as enthusiastic designers came to grips with the evolution of a truly effective firing mechanism (with trigger-operated S-shaped match holder and covered pan to prevent the priming

Overleaf:
A painting by Sebastian Vrancx of a camp scene in the Thirty Years War emphasizes the diversity of weapons in service at the time. The massive furniture of the firearms provided a sound structural basis for all the metal parts.

charge from blowing away) on a weapon with sights and with shaped wooden furniture to sit more comfortably in the firer's hands. There were many variations on the theme, but what is indisputable is that the forerunner of the true small arm had arrived, for all that the weapon was still large, heavy and relatively clumsy in use, confining its military applications to comparatively static engagements on wheeled carts and the like.

The evolution of the effective firearm, sparked by the matchlock, also resulted in a considerable growth in allied experimentation. The period from the end of the 15th century saw widespread development in Europe of rifled barrels (with helical grooving in the barrel to spin the projectile and thus improve its stability once it had left the muzzle), increasingly accurate sights, interchangeable barrels (to allow different calibers or even types of barrel with the same basic firing mechanism and furniture), breech-loading and preloaded chamber pieces for faster rates of fire, repeating weapons (generally with cylindrical magazines) and multi-barreled guns. There was much technical and operational virtue in most of these developments, but most foundered on the inability to provide gas-tight seals between the breech and the barrel, leading to loss of gas pressure with consequent loss of range and force as well as danger to the firer.

Increased experience, design skill and manufacturing capability also played a great part in reducing the size and weight of small arms, with the result that the pistol became

Below:
An early 17th-century illustration details just six of the 36 movements needed to load, aim and fire a matchlock musket of the period—a time-consuming activity that had to be perfected.

Right:
This 19th-century engraving provides interesting detail of early small arms types from the matchlock to the percussion lock, indicating that not all weapons were intended for military use during times when field sports were very popular.

Below:
With the development of early firearms as practical weapons, the art of the gunpowder manufacturer became increasingly important; this manuscript illustration of the late 14th century shows stages in the making of gunpowder in a German factory.

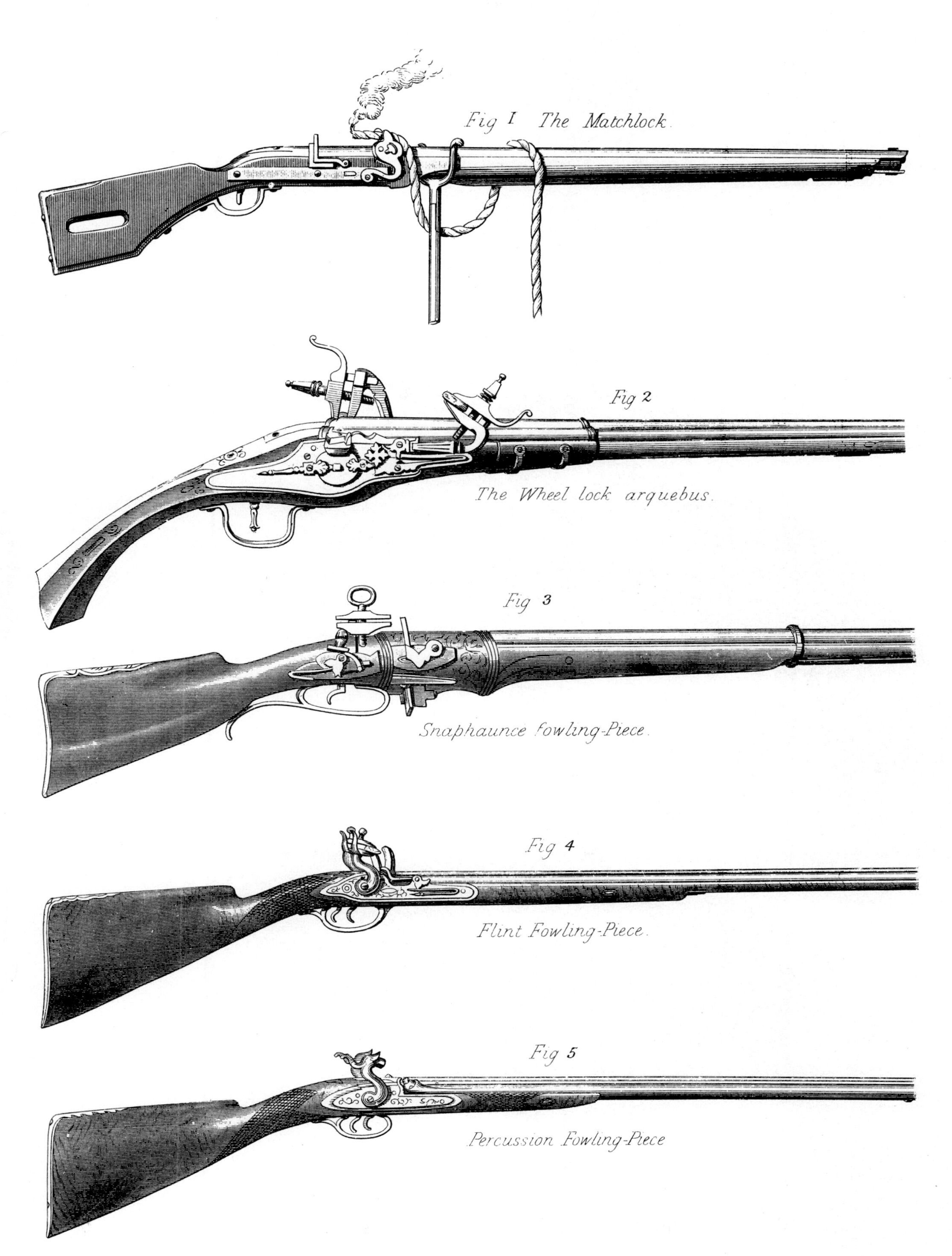

Right:
A print of 1735, The Granadiers [sic] Exercise at the Granade *reveals the stiff formality of musket drill in the period, when uniformity of volley fire with indifferently accurate weapons was far more important than individual musketry skills. The weapon is a typical flintlock of the period.*

Below:
A pair of flintlock rifled pistols made for Napoleon I by Lepage of Paris—excellent examples of the gunsmith's art at the beginning of the 19th century.

Bottom right:
The poor standard of musketry and the low rate of fire possible with muzzle-loading weapons placed great emphasis on volley fire by three or more ranks of infantry in tight squares. Such fire (with bayonets as the final line of defense) proved decisive against French cavalry at the Battle of Waterloo in 1815.

Top right:
The charge of the Life Guards in the Battle of Waterloo. In such engagements the shock of the cavalry charge was greatly heightened by the discharge of pistols and carbines just before the charge struck home, whereupon the sabre came into its own as a far more practical close-quarters weapon.

relatively common, and the larger small arms far more mobile, to the detriment of the armored nobility, who lost their military superiority which had been based on protection and mobility. From this time forward, therefore, the gun-armed infantryman may be judged one of the arbiters of the battlefield even though cavalry still had a major role to play.

The wheel-lock

For all that it was one of the great inventions of military history, the matchlock had one great failing: the match burned away, was dislodged from the jaws of the lock or was extinguished by rain. Gun designers were fully aware of this limitation, and thus sought an alternative method of igniting the priming charge. The result was the wheel-lock, probably invented in the first quarter of the 16th century in Germany or Austria. The principle is simple: instead of the match and its holder, the wheel-lock had a rough steel plate that was turned by a key-operated pre-wound spring against a flint when the trigger was pulled, the resultant shower of sparks falling on the priming charge and igniting it. The wheel-lock was adopted with great alacrity all over Europe, and appeared in countless forms. One side effect of the type's adoption was the development for the first time of safety mechanisms: when small arms had been difficult to fire there seemed little point to such devices, but the simplicity and consistency with which a wheel-lock could be fired prompted the adoption of devices to ensure that the weapon could not be discharged accidentally.

The flintlock

One disadvantage to the wheel-lock, however, was its high price. This led to the snaphaunce, an updated version of the matchlock with pyrites held in the jaws of the lock and designed to snap down on a steel anvil beside the priming charge and so produce the necessary shower of ignition sparks. The snaphaunce first appeared in about 1525, but it took about a century before the logical development of the type appeared in the form of the classic flintlock. This worked in much the same way as the snaphaunce, but was arranged so that the downward movement of the cock towards the steel lifted the frizzen otherwise covering the priming pan to prevent the priming powder from blowing away or being wetted by rain. It was only from about 1625 that the flintlock began to gain universal acceptance, and it rapidly became the classic action for muzzle-loading small arms during the second half of the 17th century.

With the reliable action of the flintlock, designers and gunmakers devoted very considerable efforts to perfecting the type. Improved powder and manufacturing processes played an important part in the rapid progress made with flintlock pistols and muskets in place of the earlier arquebuses, while better iron alloys superseded bronze and brass as small arms mediums, making the development of lighter, stronger and more accurate weapons possible. As with the matchlock, designers were extremely bold in the development of variations on the theme, mostly designed to produce a higher rate of fire. Such experiments and a few 'production' examples centered on breech-loading using an adequately obturated opening breech mechanism for the rapid loading of single rounds, or alternatively attempting to combine a revolving magazine system and semi-automatic priming system. Enormous ingenuity was employed in these systems, but the relatively inaccurate manufacturing processes of the time were largely responsible for such experiments remaining just that. It is worth noting, however, that the first primitive mass-production techniques provided service weapons with a modicum of interchangeable parts, greatly easing the logistics aspect of weapon maintenance, and also markedly improving tactical reliability.

Napoleon decorates one of his soldiers in 1811. This illustration serves to highlight the importance attached to perfect dressing and musketry drill for the infantry of the period. Only thus could a totally united face be shown to the otherwise overwhelming arme blanche, *the cavalry.*

Right:
Scots infantry of the 11th (71st) Regiment at the Battle of Vimiera reveal the length of the flintlock of the time, complete with the long bayonet that en masse provided a formidable hedge against enemy cavalry.

Below:
An example from a three-rank firing diagram of 1790 shows how carefully this tactic was undertaken by the British. The signals of the officers and NCOS were of paramount importance as shouted commands were often inaudible in the brouhaha of battle.

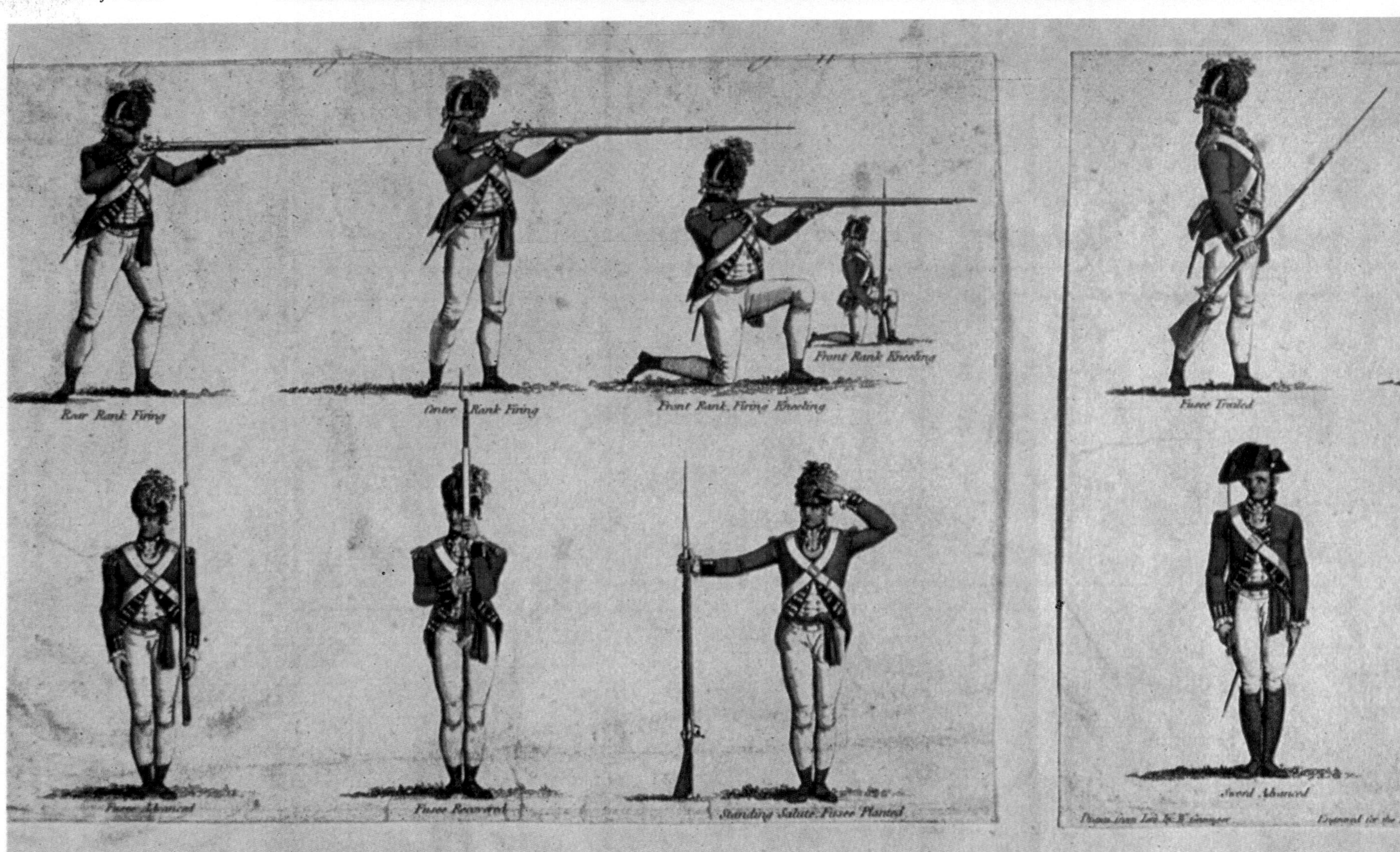

Left:
British riflemen of the 60th and 95th Regiments; the private of the 60th on the left is shown priming the pan of his weapon in preparation to fire.

Right:
A private of the Royal Marines in 1815. These men were well versed in the arts of marksmanship, and were used on board Royal Navy warships to fire from the fighting tops on the masts on the enemy's officers and petty officers in any close engagement.

Below:
Belgian burghers surrender the keys to their city as seen in a manuscript illustration from the end of the 15th century. In the background the victorious besiegers break camp while the cannoneers' devastating weapon may well have helped to create breaches in the walls.

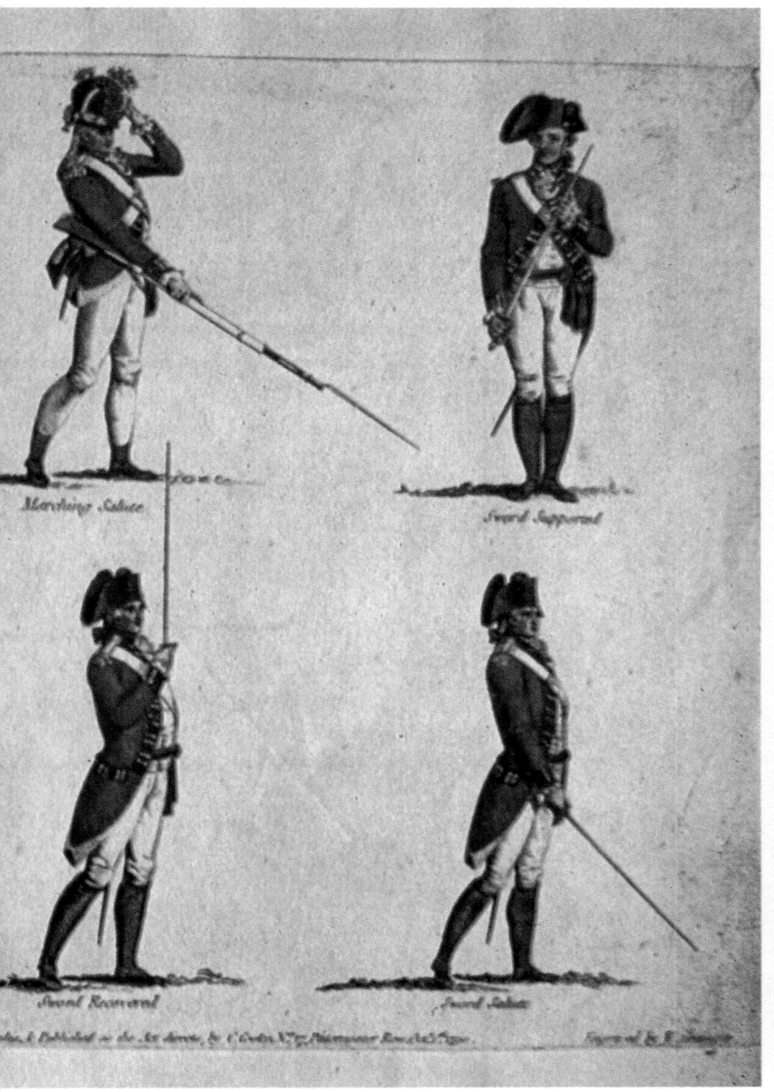

Left:
The tactical mobility of the matchlock was low because of its weight, so the use of a vehicle like this German war chariot made a world of difference to the firer, who relied on more conventionally-armed soldiers to protect him while he aimed, fired and reloaded.

Below:
An engraving of Louis XIV's musketeers at practice. Note the rest needed to steady this heavy weapon, and the match used by the musketeer on the right (one end in the lock and the other free for emergency use).

Rifles

This improved manufacturing capability also made possible the evolution of the rifle as an effective weapon. With its rifled bore imparting considerable spin to a lead ball, which thus had a certain gyroscopic stability in the air and hence improved accuracy, the rifle was more expensive than the musket, and so lacked attractions for the military rapidly becoming accustomed to the advantages of mass-produced standardization. The military also appreciated that the musket, with shorter barrel and looser-fitting ball, was easier and thus quicker to reload, allowing ranks of infantry to deliver devastating volleys of fire at short ranges and over fairly protracted periods. The rifle appealed to specialists, however, for it had far greater accuracy and range. The value of the weapon was amply proved by American irregulars in the War of Independence, when their ability to blend with the country and pick off British troops at long range while also securing valuable tactical reconnaissance proved to be one of the most important military lessons of the period. From this period onwards, therefore, the conventional military virtue of concentrated but short-range musket volleys was at first complemented and then ultimately replaced by rifle fire, which was used initially by special units trained for reconnaissance and for harassment with accurate long-range individual fire.

So far, however, the development of small arms had been limited by the use of gunpowder as the propellant and in the main charge's ignition system. Even with the adoption of the frizzen-equipped flintlock such a system had severe limitations on tactical use of the weapons, for the priming was still vulnerable to an extent to any damp, and could be blown as the frizzen opened. However, gunpowder had to be ignited to initiate the firing process, and as long as gunpowder remained the only suitable primer the gunmakers were stymied.

Above:
In all essential respects, the arquebus of the early 16th century was a little cannon, massively stocked to turn it into a man-portable weapon. Given the temperability and complex reloading of these weapons, a sword was an essential item of back-up weaponry.

Below:
Those who could afford arquebuses were usually men of some substance who would delight in seeing themselves presented with a measure of swagger, as indicated by this engraving from a treatise by Jacob de Gheyn showing Lowland arquebusiers of the early 17th century.

Above:
Trappers fighting off Indians with muzzle-loaded flintlocks, by F. Remington.

Below:
American Hunting Scenes: A Good Chance *by Currier and Ives in 1863, shows frontiersmen in pursuit of game. The rifle was all-important in such situations giving its owner a far better chance of bringing down his quarry than an unrifled weapon; success with the rifle meant survival for the firer and his party.*

Right:
Some of the earliest exponents of the rifled musket were the American woodsmen of Kentucky and Tennessee, who adopted the type from German immigrants and turned what was originally a hunting weapon into a formidable firearm for hunting and the frontier warfare peculiar to that time. This painting by Gayle Hoskins entitled Trade from the Monongahela *shows aspects of rifle manufacture, repair and general maintenance.*

17 PHLF 36

From the percussion cap to nitro-cellulose

Detonation developments

It was at this point that the chemist entered the scene to remove the gunmakers' reliance on gunpowder as the priming agent, the essential link between the mechanical action of the lock and the chemical transformation of the main propellant charge from solid grains to a rapidly expanding volume of hot gases. What was needed was a means of converting the mechanical energy of the lock into chemical energy that could be used to fire the propellant charge. Such a means was found initially with fulminate of mercury, which explodes when hit sharply. This had been discovered at the very beginning of the 18th century, but it was not until the start of the next century that the principle was used in weapons, initially in an adaptation of the basic flintlock action: powdered fulminate of mercury was used instead of the gunpowder priming, the shock of the cock hitting the pan detonating the fulminate of mercury and hence the main propellant charge. The primary figure in this development was a Scottish minister, Alexander Forsyth, who developed a practical percussion pistol in 1805, and then percussion caps as an advance on loose powder. In the cap system the fulminate of mercury was sandwiched between two layers of thin paper that could then be produced in a long roll. The development of the new lock was rapid, and the paper percussion cap was rapidly supplemented by a tubular variety and, in 1814 or thereabouts, by the definitive copper cap, in which the fulminate of mercury was enclosed in a copper shell with a short skirt that would fit over a nipple extending from the touch-hole for detonation (and subsequent ignition of the priming powder under it) as the hammer of the lock fell.

The percussion lock eliminated the ignition uncertainties of the flintlock and its predecessors, and rapid strides in automating or semi-automating the loading followed.

Below:
French Imperial Guard Zouaves during the Crimean War, in a pose of studied nonchalance with their long-pattern pillar-breech percussion rifles.

Left:
In remote areas obsolescent weapons such as these percussion locks remained viable right into this century, their technological obsolescence being more than offset by familiarity, reliability, the possibility of local repair and ammunition manufacture—and the lack of more advanced weapons in the hands of opposing forces.

Bullet developments

From the 1830s onwards the rifle rapidly displaced the musket, the Napoleonic War having shown conclusively that the rifle was the better weapon. The practice of the time was to make the ball very slightly smaller than the bore of the rifle so that when wrapped in cloth it could be rammed down the barrel for a tight and relatively gas-proof fit. The next step was to produce a ball with protrusions that would fit into the grooves, which at first numbered just two. By the early 1840s it was appreciated that the spherical ball was inferior ballistically to a conical type, and the first cone-nosed bullets were issued in the mid-1840s despite problems with barrel-fouling.

The solution, at least in the short term, was provided by a Frenchman, Capitaine Charles Minié, who developed (perhaps as a result of researches into similar work by the British W. Greener) a slightly elongated conical bullet (with a diameter slightly less than the bore of the rifle to ease loading) whose hollow base was fitted with a small iron cup: as the powder in the firing chamber was detonated, the iron cup was driven up into the base of the bullet and caused it to expand across the full bore of the rifle, preventing the propellant gas from escaping and thus providing maximum muzzle velocity, with minimal barrel fouling. There were still problems, mainly because of the poor standards of manufacture in barrel and bullet, and though a number of experimental solutions were tried (including a hexagonal bullet in a hexagonal barrel), the solution was finally found to lie in more accurate manufacture, largely with the aid of machine tools from the rapidly developing US industrial base.

Conical bullets in rifled barrels were accurate and copper percussion caps were reliable, but muzzle loading kept the practical rate of fire far too low. Improvements had been made with adoption of the paper cartridge, a waxed or greased container for the bullet and just the right quantity of powder for a single shot: the firer tore open the cartridge, poured the powder into the barrel, and then rammed the cartridge paper down as a wad before spitting in the bullet and ramming it down. But he still had to place his weapon butt down on the ground, get out the cartridge, go through the loading motions, prime the weapon and fit a cap before he could bring the rifle to his shoulder again.

Breech loading

The solution was clearly breech loading, which would remove the need to ground the butt and use a ramrod. An extraordinary number of such systems were evaluated during the second quarter of the 19th century, and most foundered on lack of obturation: no matter how tightly closed the breech seemed to be, firing a round inevitably resulted in massive loss of propellant gas through the breech, endangering the firer and losing gas for the propulsion of the bullet. Slowly the problems were solved (in part at least), and during the mid-1800s there were four practical breech-loading systems in service. The most famous of these is the bolt action invented by a German, Nicholas Dreyse, for his Zündnadelgewehr (needle gun): in this a sliding bolt gave access to the breech, into which a paper cartridge was loaded before the bolt was closed. The cartridge contained, from front to rear, the bullet, the primer/percussion cap and the propellant powder: when the trigger was pulled, a long needle inside the bolt was driven forward through the powder to strike the percussion cap, so setting off the propellant which, Dreyse claimed, burned more evenly when ignited at the front. Though little appreciated at the time, the importance of the Dreyse system was not the needle or the cartridge, but the turn-bolt locking system, which was destined to come into its own only with the adoption of the metal cartridge later in the century.

Next in importance was the dropping-block locking system, which was favored to an extent in Europe, but far more so in the USA as the actions of the classic Sharps weapons. The system is essentially simple: lowering the hinged trigger guard pulled down the breech-block into mortises let into the frame, allowing a cartridge to be pushed into the chamber, whereupon the trigger guard was raised, lifting the breech-block and locking the rear of the chamber. In the days of the paper cartridge, raising the breech-block sheared the back of the cartridge and provided a better-than-average seal in this vital area, and removing the cartridge rear allowed the blast to the percussion cap to ignite the main propellant charge without an intervening primer charge. This further simplified operation of the weapon, and also speeded the loading process.

There were also a number of hinge-frame weapons similar in concept to the shotgun, with the barrel broken to provide access to the chamber, a good seal then being achieved by the close fit of the component parts that could now be machined to close tolerances.

The last major type, which was never widely adopted, was the revolver type generally accepted as originating with Samuel Colt in 1836. In this system, later to become univer-

Far left, top:
This camel is in for a highly unpleasant shock if his rider, an Indian camel cavalryman, fires his flintlock rifle.

Far left, bottom:
Hodgson's Horse in action against mutinous sepoys during the Indian Mutiny of 1857, a grisly war sparked by Indian fears that the cartridge for the new Enfield rifle was waterproofed with substances that would defile either Moslems or Hindus.

Below:
Men of the 13th Light Dragoons display short carbines in this 1853 illustration. These weapons gave cavalry a useful firepower punch as they charged, but they had to be short enough to be stowed in a boot alongside the saddle and handled with one hand during the charge. The result was a light weapon with minimal accuracy, a factor of little importance at close range.

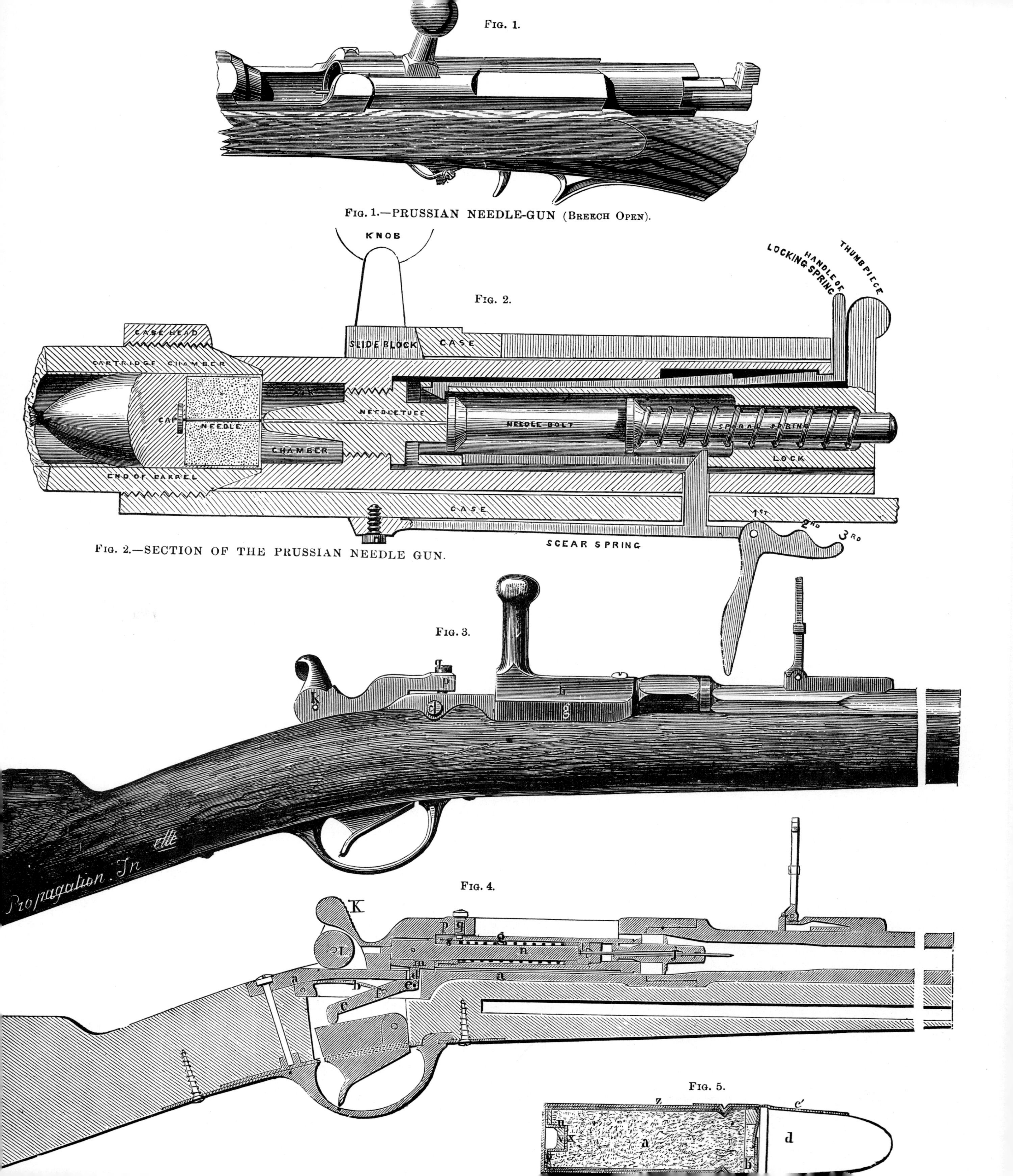

Fig. 1.—PRUSSIAN NEEDLE-GUN (Breech Open).

Fig. 2.—SECTION OF THE PRUSSIAN NEEDLE GUN.

sal on revolver pistols, the cylindrical magazine was loaded with five rounds that were brought into line with the barrel by a mechanical action derived from movement of the trigger. As with all breech-loading weapons before the advent of fixed ammunition the Colt had distinct limitations, but it did offer the priceless tactical advantage of ready-to-fire rounds when the opposition (unless he too had a Colt or other revolving rifle) had only a single round. So even though the Colt revolving rifle had its failings, it pointed the way forward for military weapons with its multi-shot capability through the use of a magazine and mechanical reloading.

The second quarter of the 19th century is thus of intense interest as the time in which the basic ingredients of the modern rifle and handgun began to appear in acceptable form. No weapon yet had all the requisite features, but several had begun to pull together diverse features that were contributing towards far more satisfactory weapons. The limitations now were lack of impetus (there were few full-scale, protracted wars in the period to stimulate development, or even test current weapons) and the failure of ammunition development to match weapon development.

Fixed ammunition

However, the success of the percussion cap persuaded designers that it was now possible to produce fixed ammunition (a case containing cap, primer and propellant, and tipped with the bullet) for direct loading into the breech. Although a number of extraordinary ammunition designs were produced, evaluated and rapidly consigned to the historical scrap heap, the very proliferation of fixed ammunition designs is clear proof that the basic virtues of the type had been amply recognized: such ammunition was easy to carry and load, the latter being so straightforward that a great increase in rates of fire could be expected. The needle-fire round of Dreyse appeared in 1830, while other interesting rounds of the period were tit-fire, in which a copper-cased nipple containing the cap protruded through a hole in the rear of the breech-block, to be struck and detonated by the falling hammer, the Demondion tail-fire in which opening the rifle for loading compressed a spring that, once the rifle had been loaded and the breech closed, snapped up on operation of the trigger to crush against the underside of the block, and so detonate a flat percussion tail projecting from the rear of the cartridge, and the pin-fire, developed from the Demondion but using a pin driven by the hammer to strike and detonate the cap.

Perhaps the most important of the pin-fire types was the Lefaucheux cartridge, which appeared in France during 1836. Although the pin-fire mechanism was eventually abandoned as unnecessarily complex and prone to accidental damage, the Lefaucheux cartridge for the first time

Left:
The workings of the Dreyse Zündnadelgewehr *(needle gun), the world's first bolt-action breech-loading rifle. This was an epoch-making weapon whose technical details and performance the Prussians tried desperately to keep secret, though it was the overconfidence of other countries that actually prevented them overtaking the Prussians in the next 30 years.*

Right:
An NCO of the Union army's engineer corps during the American Civil War poses with his Model 1861 rifle, a highly accurate muzzle-loading weapon fitted with a fearsome bayonet. A skilled rifleman could fire off as many as 10 rounds while the enemy advanced the typical engagement range of the Civil War. Leaning against the drums is a Spencer repeater.

successfully addressed the problem of breech obturation to prevent gas escaping. The Lefaucheux cartridge had a cardboard body, but a brass base that expanded under the pressure of the propellant gas to seal the gap between the chamber and the breech-block, so preventing the gas from escaping except up the barrel behind the bullet. Lefaucheux also produced a hinge-frame gun to fire his new cartridge, and the combination must be regarded as of great historical importance as the true progenitor of modern shotguns and cartridges.

Such were the weapons and the cartridges that began to find more common use in the second quarter of the 19th century. They were generally more satisfactory than the muzzle-loading muskets and less numerous rifles that had reigned supreme up to the end of the Napoleonic War in 1815, but were generally underdeveloped and all too prone to fail under the stress of combat. Nevertheless, the Dreyse gun gave the Prussians decisive advantages in the Prusso-Danish War of 1864 and the Prusso-Austrian War of 1866, for despite its many failings and an effective range as short as 150 yards the Dreyse breech-loader was a quantum step forward over the muzzle-loaders still used by the Danes and Austro-Hungarians. The French from 1866 used the similar needle-fire Chassepot, with a rubber seal which eventually turned into the consistency of rock with the repeated heat and pressure cycles of firing, and the British were experimenting with ways to convert their Enfield muzzle-loaders into breech-loaders with a massive competition eventually won by the American Jacob Snider. In the Snider conversion the breech was cut open and the resulting breech-block was hinged to open to the right and fitted with a firing pin and spring operated by the original lock. One of the best features of the Snider system was a claw on the breech-block to extract the spent cartridge, but a disadvantage was the need to turn the rifle over to make the hot cartridge case fall out.

By this time, it should be noted, the Americans were far ahead of their European counterparts, for the Sharps and Henry falling-block weapons were already in widespread service, and were undoubtedly far superior to those used by the Europeans. Both were completely reliable and very strong, and Henry's pioneering work with metallic cartridges and lever-action repeating rifles led to great improvements in the near future. It should be noted, however, that the starting point for Henry's work was the impractical but visionary Hunt Volitional Repeater of 1849, operated by two levers, one for the rocket-type ammunition and the other for the separately-fed primers. The basic concept of the Volitional Repeater was then the starting point for Lewis Jennings, who evolved a repeating rifle with two features that were to become standard: the tubular magazine under the barrel and the toggle-lock action. Henry was involved in the making of the Jennings prototype for Robbins and Lawrence of Windsor, Connecticut, and then moved to Smith & Wesson, for whom he

The American Civil War was fought with an extraordinary assortment of firearms ancient and modern, ranging from War of Independence muskets to the very latest repeating rifles. This is the battle of Kenesaw Mountain in 1864, with remarkably well-equipped and well-uniformed Confederates apparently repulsing a Union assault.

worked on the Volcanic repeating rifle, combining aspects of the Jennings rifle with a special type of ammunition (each round was hollow-based to contain its own limited quantity of propellant powder held in place by a card base containing the priming charge) and the magazine for separate 'percussion pills' designed by Horace Smith. The Volcanic was not a success, but paved the way for Henry's later work.

Cartridges

The solution to the problems of obturation now besetting the weapon designers lay with the metallic cartridge. This could contain the powder, primer and percussion cap, support the bullet at its front, and yet be made of a material ductile enough to expand readily as the round was fired and so seal the gap between the cartridge and the inner wall of the chamber to prevent the rearward escape of gases, then contract as the gas pressure fell to allow the spent case to be extracted easily. Another key requirement of the metallic cartridge, and one often overlooked, is for it to provide the percussion cap with a solid "anvil" on which it can be crushed by the hammer to ensure reliable detonation. The way had already been shown by the Lefaucheux pin-fire cartridge, which reached a practical level of reliability only when the modifications of another Frenchman, Houiller, had been incorporated. In such cartridges, designed primarily for sporting guns, the lessons of adequate obturation were evident, but most military authorities were content to leave the development of effective cartridges to the civilian market in Europe and to the Americans. The Americans drew the correct conclusions from the scientific evidence, which was firmly backed by an examination of muzzle-loaders abandoned at the Battle of Gettysburg in 1863, when muzzle-loaders were still the most numerous weapon type in the armories of both sides, even if they were percussion-fired rifles. The survey provided clear proof of the problems associated with muzzle-loading weapons in the heat of combat, and the US authorities thus concentrated on breech-loaders using the most effective cartridges available.

The next step in the development of metallic cartridges was the rim-fire cartridge, whose basic principles are inherent in much of Houiller's work and in the light cartridge produced in France during the early 1830s by Flobert for pistols and what are popularly termed "parlor rifles" of very low power. However, it was the Americans who now

Below:
A trapper loads his plains rifle at the gallop in this illustration by F. Remington, who has given the rifle a remarkably short barrel for such a weapon!

took the lead, and the first successful rim-fire cartridge was produced in 1857 by Smith & Wesson. The company was already in the forefront with its process of producing comparatively cheap drawn cartridge cases, and it was thus straightforward to include a priming charge spun into the rim projecting outwards from the body of the case. This rim prevented the round slipping too deeply into the firing chamber by coming to rest against the base of the chamber, which thus served as an admirable "anvil" against which the primer-laden rim could be struck by the hammer, ensuring reliable detonation.

A brass or copper case also solved the problem of obturation, as noted above, which allowed the development of smaller and handier rounds, in turn leading to smaller and lighter weapons. Loss of propellant gas in inadequately-obturated weapons such as the Chassepot and weapons of its ilk meant that power was low, and to compensate a fairly massive bullet was used: the Chassepot fired an 11 mm/0.434-inch round and the Enfield a 0.577-inch/14.66 mm round, both inefficiently. The successful development of the Smith & Wesson rim-fire cartridge with ductile case meant that rounds as small as 0.22-inch/5.6 mm became feasible for pistols and sporting rifles, though many military weapons were still chambered for calibers as large as 0.44-inch/11.2 mm, or 0.45-inch/11.4 mm because of the limitations of black-powder propellants and shorter barrels. However, the round fired even by the large-caliber weapon was smaller than that used in older guns, the smaller mass being more than balanced by considerably greater muzzle velocity for improved kinetic energy.

Ammunition reached its definitive form with the evolution of the center-fire cartridge as an alternative to the rim-fire type, which is still used in a number of handguns. The center-fire type appealed to the military in particular as it made reloadable rimless ammunition possible (the distortion of the rim-fire type's outer edge made reloading impossible) in the increasingly popular bolt-action weapon. This had a firing pin moving axially through the center of the bolt to strike the percussion cap centered on a depression in the base of the cartridge. The first effective center-fire round was produced in France during 1857 by Pottet, but it was nearer to the card-cased shotgun type than higher-powered rounds, and the first practical center-fire was the one introduced in 1861 by another Frenchman, F. Schneider. It was this type that was specified for the British Snider-Enfield rifle produced after the Prusso-Danish War

Below:
A dead Confederate sniper on the field of Gettysburg during the American Civil War, during which several battles were fought in fixed trench lines. Sniping with weapons such as this Springfield Model 1861 rifle was therefore an important military skill to keep the enemy off balance and, if possible, to pick off enemy officers.

Left:
A classic weapon of its type, and a mainstay of the British army during the Crimean War: the Enfield rifle in 0.577-inch caliber, here dated 1853, and the percussion lock, 1854. Under the barrel, as with all such weapons, is the vital ramrod. In the army of Frederick the Great, the ramrod was considered so important that its loss sometimes carried the death penalty.

Below:
Caught unprepared in the Battle of Isandhlwana by the Zulus during 1879, the British could not use their Martini-Henry rifles to advantage and were annihilated in close-quarter fighting—a clear enough indication that long rifles had their limitations.

Bottom right:
The 0.577-inch waxed paper cartridge for the 1853 Enfield rifle, allowing comparatively rapid loading with exactly the right amount of powder.

had confirmed the overwhelming superiority of the breech-loader. But now all the ingredients were there: the metallic cartridge provided good obturation and contained the right quantity of propellant, ignited with increasing reliability by an inbuilt primer, to propel a conical bullet mounted at the front of the case.

The availability of strong and reliable metal-cased ammunition opened the way for more effective firearms based on breech-loading mechanisms, generally of the four types mentioned above and initially of the single-shot type as breech-loading increased the practical rate of fire of muzzle-loading weapons dramatically.

Breech-locking dropping-block

A problem with breech-loading is the very strength needed in the system for locking the moving part of the system to the fixed breech and barrel safely and reliably. The earliest and still the strongest breech-locking mechanism is the dropping-block system designed by Christian Sharps during the period of the percussion cap. Its enormous strength and unexcelled obturation were even more important with the adoption of the metallic cartridge. As a result most of the original Sharps weapons were converted for rim-fire and center-fire metallic cartridges, and the type maintained a superb record of reliability and accuracy. The Sharps rifle in varying calibers was the preferred weapon of buffalo hunters and "sharpshooters". Other dropping-block weapons of the third quarter of the 19th century were generally American, typical being the Sims, the Stevens, the Ballards and a 0.22-inch Winchester designed as his first successful exercise by John M. Browning, undoubtedly the greatest single figure in the history of small arms design.

The falling-block system attributed to Henry Peabody is similar in concept. It was introduced in 1862, and has a rear-hinged breech-block on a line higher than the center of the bore. When the trigger guard is pushed down and forward, a lever pulls down the leading edge of the block, opening the rear of the chamber and extracting the spent case so that a fresh round can be loaded before the raising of the trigger guard closes the action. There was some US interest in the Peabody system, but the end of the American Civil War removed the spur for a radical innovation to be accepted, and the falling-block system thus found favor initially with the Europeans, who adopted it widely and also ordered weapons directly from Peabody after exhaustive tests into reliability, safety and accuracy. A Swiss engineer, Frederick von Martini, revised the Peabody action (with its external and separate hammer) to include initially a mechanism, worked by the loading action of a lever behind the trigger group, to cock the external hammer, and then a hammer inside the breech-block itself. The Martini system was accepted by the British army in 1871 for the classic Martini-Henry rifle, which combined the Martini action with the polygonally-rifled barrel designed by Henry. Other notable European manufacturers and designers to adopt and then adapt the Peabody falling-block lock were the Bavarian Werder, the Belgian Francotte-Martini, and the British Westley-Richards.

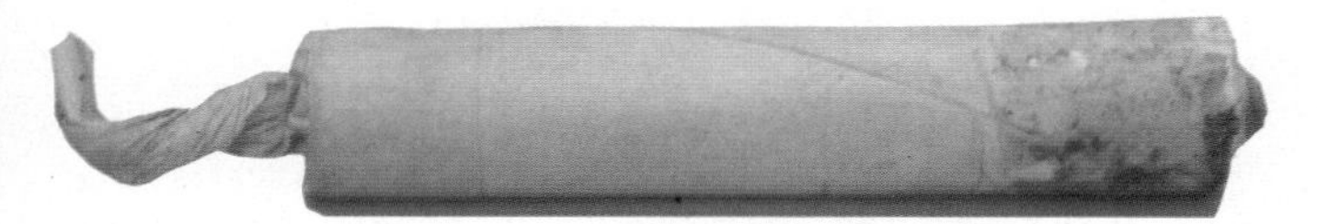

Rolling-block

The rolling-block system is similar in concept. Devised by an American, Leonard Geiger, in 1863, it is an immensely strong yet fully effective system. It uses two major moving parts, the heavy breech-block and the lighter hammer, each mounted on its own transverse pin below the level of the chamber. The breech-block is fitted with a spur so that a thumb can roll the block back and down (against the pressure of a spring that otherwise keeps the block against the face of the breech) once it has been unlocked, and in the center of the rolling-block is a hole containing the firing pin. As the trigger is pulled and the hammer moves forward, a projecting platform slides below the undersurface of the breech-block and locks it in place so that the hammer can hit the firing pin only with the breech locked shut. A similar design was evolved by Joseph Ryder with the hammer's axis in front of the breech-block's, and the two systems were combined by Ryder into the Remington-Ryder system after the Remington company had bought both patents. By 1866 Remington was ready for production, but like the Peabody lock the type found favor with European forces rather than with the US military, and useful sales were also made in South America and China.

Rifling

But whereas the Americans proved the most inventive with the block-type actions for single-shot rifles and the Germans with the bolt-action rifle, the main advance with rifling came from the UK. Rifling had long been appreciated as a major contributor to accuracy, especially at longer ranges, but more consistent propellants and metallic cartridges made effective rifling imperative to make full use of weapons' greater and more reliable range. Two keys to the development of effective rifling were a better understanding of ballistics, and the machine tools (initially of US manufacture, but after 1854 slowly replaced by British equipment) to produce accurate rifling.

Up to about 1860 rifling had been polygonal, which though effective was particularly prone to black-powder fouling of the grooves in any sustained engagement. But in 1865 the British engineer Metford devised a new type of shallow rifling with curved rather than angled grooves and lands. The type was an immediate success, for it appreciably reduced fouling, improved accuracy and reduced the distortion that had previously afflicted both bullet and barrel. It is also worth noting that it was the Metford rifling system that introduced the 0.303-inch/7.7 mm caliber, which rapidly became the 'classic' British rifle caliber.

Metford rifling soon became standard throughout the world, and was supplanted from the 1880s by the Enfield type only because of the introduction of nitrocellulose propellants, which eroded the smooth contours of the Metford rifling far more rapidly than black powder had. The Enfield rifling was developed specifically for use with the new 'smokeless' propellants, and has deeper square-sided grooves concentric with the bore of the barrel.

Bolt action

Hand-in-hand with the new type of rifling devised by Metford was the classic bolt action invented by the only real competitor to John Browning as the single most important small arms designer, the German Peter Paul Mauser. Trained at the government factory in Oberndorf, Mauser was well aware of the advantages of the Dreyse gun's adequate but primitive bolt action, which gave the Germans a decisive tactical edge in the Prusso-Danish and Prusso-Austrian Wars. After a period in the army Mauser

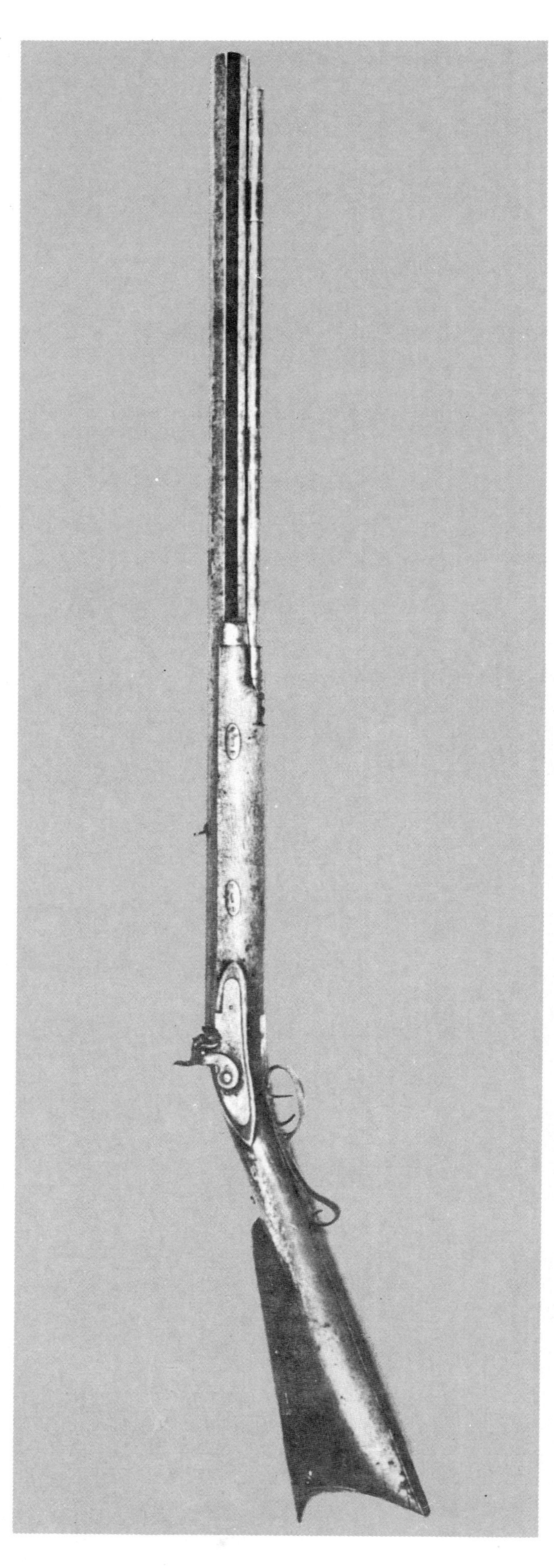

Left:
A classic of its type: a plains rifle made by Jacob and Samuel Hawken and owned by the legendary Jim Bridger. The extremely strong octagonal barrel was essential in a weapon designed for serviceable use in the remote areas of the American plains.

Right:
This photograph of a lieutenant and two privates of the 4th Michigan Infantry, Union army, indicates the importance rightly attached by such infantrymen to their weapons. Though the private on the left is somewhat self-conscious with his Colt revolver, the man in the center holds his Minié-type rifle with the ease of long association.

Overleaf:
The revolver pistol in its current form is largely an American development. Clockwise from top left, these important early weapons are the Colt Dragoon Model 1849, the Remington Army Model 1863, the Starr Model 1856 revolver and the Colt Navy Model 1851.

STARR'S PATENT JAN. 15. 1856.

Left:
The real face of the American West: a Winchester rifle and Colt revolvers.

Below:
Gurkha riflemen of the British army drill with the Snider-Enfield, a short-term expedient evolved from the Enfield to provide an interim breech-loader while a proper breech-loader was developed.

decided to investigate the possibilities of a better bolt action using the new type of metallic cartridge, about which he received much information from his brother Franz, who had emigrated to the USA and worked for Remington. There had already been several largely unsuccessful attempts to improve on the Dreyse action, but Mauser proceeded with a single-minded dedication to solve a host of related problems. The first step was the invention of an automatic cam-cocking system in which the firing pin was withdrawn as the bolt was opened, eliminating accidental discharge as the bolt was closed, and also making possible the introduction of effective extraction and ejection of the spent cartridge. Mauser then moved on to the design of the bolthead face, which was made quite gas-proof and made removable, and of the classic Mauser turn-bolt locking system, which offers unexcelled strength and safety. Mauser was unable to interest the German authorities in his work, and thus secured his initial patents in the USA with the aid of the European representative of Remington. Plans for production in the USA and Belgium fell through, but the associated political furore persuaded the German army to re-examine Mauser's work, resulting in the Mauser Gewehr 71 production model. Though further vicissitudes were to attend Mauser's endeavours, from this time forward he was accepted as Germany's premier rifle designer, and as such was able to work consistently towards the development of improved models which really came into their own after the introduction of "smokeless" propellants in the late 1880s.

Repeating weapons

With effective bolt and moving-block breech-loading rifles available, and the French decisively defeated in the Franco-Prussian War of 1870-71 by an emergent Prussia whose forces used in the main the obsolescent (if not obsolete) Dreyse needle gun, European powers now started a mad scramble towards new and better rifles. The next logical step was a rifle in which a fresh round could be loaded directly into the chamber from a preloaded magazine by the weapon's breech-block action. Only thus, it was correctly reasoned, could armies generate more firepower than opponents armed with manually-loaded single-shot rifles. And as is generally the case with such matters, inventors had been at work for some time before any official requirement was issued.

The first successful repeating rifle was the American Spencer design of 1860, which was made possible only by the development of the metallic cartridge. The Spencer looked like a conventional lever-action single-shot rifle, but in the buttstock was a tube holding some seven rounds pushed up from the butt plate by a spring. The action worked by the trigger guard lever comprised a semi-circular breech-block that had a falling-rising component for unlocking-locking and a rotary component for ammunition feed. The work of a 20-year-old, Christopher Spencer, the rifle was not even examined by busy War Department officials, and it was only the intervention of a perspicacious doorman that secured Spencer an interview with Abraham Lincoln, who was so impressed with the weapon that he

intervened to secure orders from both the Navy and Army Departments. Spencer's company boomed, but after the end of the Civil War orders were cut back and the company was bought by Oliver Winchester, who sold off the surviving stocks of Spencer rifles and thus eliminated the only serious competitor to his own company.

Another pioneer of repeating rifles was Tyler Henry, who only a few months after Spencer patented a Winchester repeater. This took longer to load than the Spencer, but the tubular magazine under the barrel carried 12 rounds, and with a 13th in the chamber the Winchester could be fired at an amazing 25 rounds per minute. The Winchester used an adapted toggle lock, in which the action of the trigger guard lever disjointed the lock and then drew it back, this action also cocking the hammer and bringing a cartridge into line with the breech-block; the round was chambered and the action locked as the lever was raised.

A major development in the succeeding Henry Repeating Rifle Model 1860 was the 'King improvement', a spring-loaded gate in the side of the breech to allow loading here rather than at the muzzle end of the tubular magazine as on the original model. With the reorganization of the New Haven Arms Company as the Winchester Repeating Arms Company in 1866, the Model 1860 became the Winchester Model 1866.

Though not designed as a military weapon, the Winchester Model 1866 soon made its mark, and in the process spurred a great renewal of enthusiasm for repeating weapons. In the Battle of Plevna in July 1877, now little remembered but decisive at the time, the Turkish infantry armed with Peabody-Martini single-shot rifles for accurate long-range fire checked the advancing Russian masses armed with Berdan rifles and Krnka breech-loaders, and then destroyed them at close range with Winchester Model 1866 repeaters which the Turkish cavalry had given up to the infantry. Each man had 600 rounds of ammunition, and the wall of fire that the Winchesters put up was beyond the imagination and capabilities of the Russians. The lesson could not be missed, and every European army with even the smallest claim to modernity started the search for repeating rifles.

The Winchester Model 1866 was followed by the Models 1873, 1886 and 1894, and though these were again not designed as military weapons (they used a low-powered 0.44- or 0.45-inch/11.18 or 11.43 mm round) they were often used in combat when volume of fire was the paramount consideration. The same applied to a number of other American lever-action repeaters such as the Ballard, Burgess, Colt Lightning, Kennedy and Marlin. The only rifle designed to take a full-power military round was the Savage Model 1899, which used a rotary magazine under the receiver and a modified dropping-block action.

Magazines

A number of European manufacturers explored the possibilities of the lever-action rifle with a tubular magazine, but the consensus was that such a system was better suited to sporting rifles than military weapons, and the Europeans thus came to concentrate their efforts on box magazines.

Box magazines

The first such magazine was the Lee type designed by James Lee, who had been born a Scot, brought up in Canada and was living in the USA. In 1879 Lee developed his amazingly simple yet revolutionary magazine: a box centered under the bolt containing a number of rounds that

Below:
Private John Sims of the 34th Foot wins the Victoria Cross in the storming of the Redan in 1855 during the Crimean War, when the Minié rifle was used by the British. This was the last "European" war in which only muzzle-loaders were used; the Prussians already had the Dreyse breech-loader in service, and similar weapons were under development in the USA.

Bottom right:
The winning of the Victoria Cross by Captain J.A. Wood of the Bengal Native Infantry at Bushire Fort in December 1856 during the Persian Expedition of 1856–57. Though both sides used rifled muskets, the Enfield weapon and superior training of the Indian forces swayed the balance.

Right:
The Battle of Ferozeshah during the First Sikh War in 1845. Faced with a disciplined and highly capable enemy, the British relied on the shock effect of concentrated infantry after an artillery bombardment to offset their numerical inferiority. The battle was one of the most vicious ever fought in India; the British were finally victorious after several infantry assaults had been repulsed.

were pushed upward by a spring to be stripped off individually and chambered by the forward motion of the bolt after it had been pulled back to open the breech and extract the previous round. Lee drew on the work of three British designers of the later 1860s who had appreciated the military limitations of the tubular magazine, but it was Lee who made the system practical and reliable. Lee also designed his Lee US Navy Rifle to use the new magazine, but then ran into problems with his backers and moved to the Remington Arms Company, whose flagging fortunes he revived with his magazine system. It is interesting to note that the early Lee magazine had no provision for a magazine cut-off, so if the weapon were to be used for single-shot fire the magazine had to be removed. This was clearly a tactical limitation, and later Lee magazines were provided with such a cut-off.

In Europe Mauser and the Austrian Mannlicher were also at work independently but on parallel lines in efforts to produce box magazine systems. France, Germany and Russia had already adopted repeating weapons with tubular-magazined rifles as rapidly as possible after Plevna (the French Kropatschek, the German Mauser Model 1871/84 and the Russian development of the US Evans), but work continued on developing militarily suitable box magazine weapons. First into the field, oddly enough, was the UK. The British army had been inclining slowly during the first half of the 1880s towards the tubular magazine for its first repeating rifle, but learning that the Germans were seeking to replace tubular with box magazines, switched tack and started an intense investigation of box magazine types. After investigating the Lee action with a side-feed magazine, a Swiss type, and the Lee action with a Metford-rifled 0.402-inch/10.2 mm caliber barrel, the British army finally settled during 1887 on the Lee action, the Metford rifling and the 0.303-inch caliber to produce the great Lee-Metford Magazine Rifle Mk I, at first issued with a round based on compressed gunpowder. Other European box-magazined rifles were not far behind, some like the Austro-Hungarian Mannlicher including developed features such as a clip-loading mechanism so that the magazine could be reloaded rapidly from rounds held together by a clip that fell through the base of the magazine after the rounds had been inserted. But the whole tenor of military firearms was about to change with the development and virtually universal adoption of smokeless propellants.

Above are described the most important of the developments affecting the two-handed military firearm, designed to be supported against the shoulder by one hand on the buttstock (with the trigger underneath) and the other on the forestock about halfway between the bolt and the muzzle. Such weapons developed (in practical terms) from the arquebus via the musket to the breech-loading rifle and to the breech-loading carbine. The carbine was evolved as the cavalry's equivalent to the rifle, with a shorter barrel of about 22 inches/560 mm. With the growing logistical com-

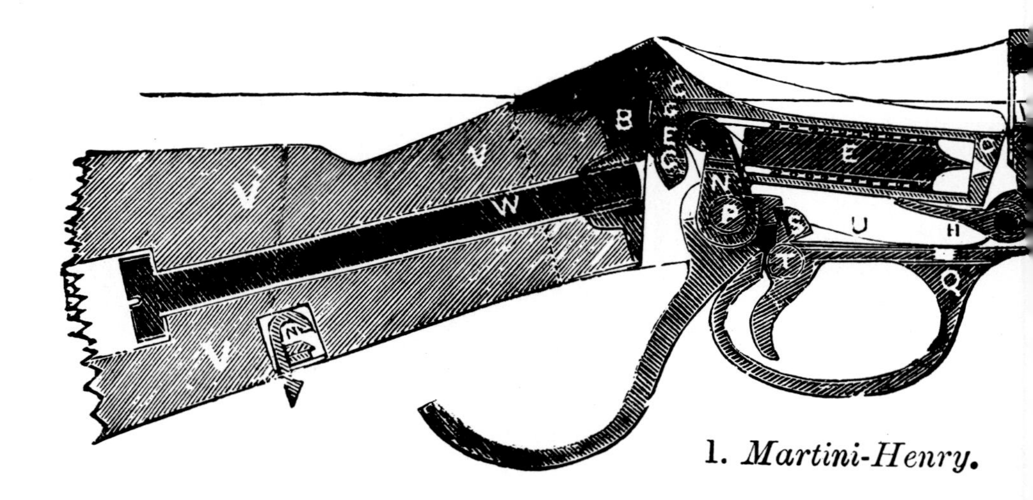

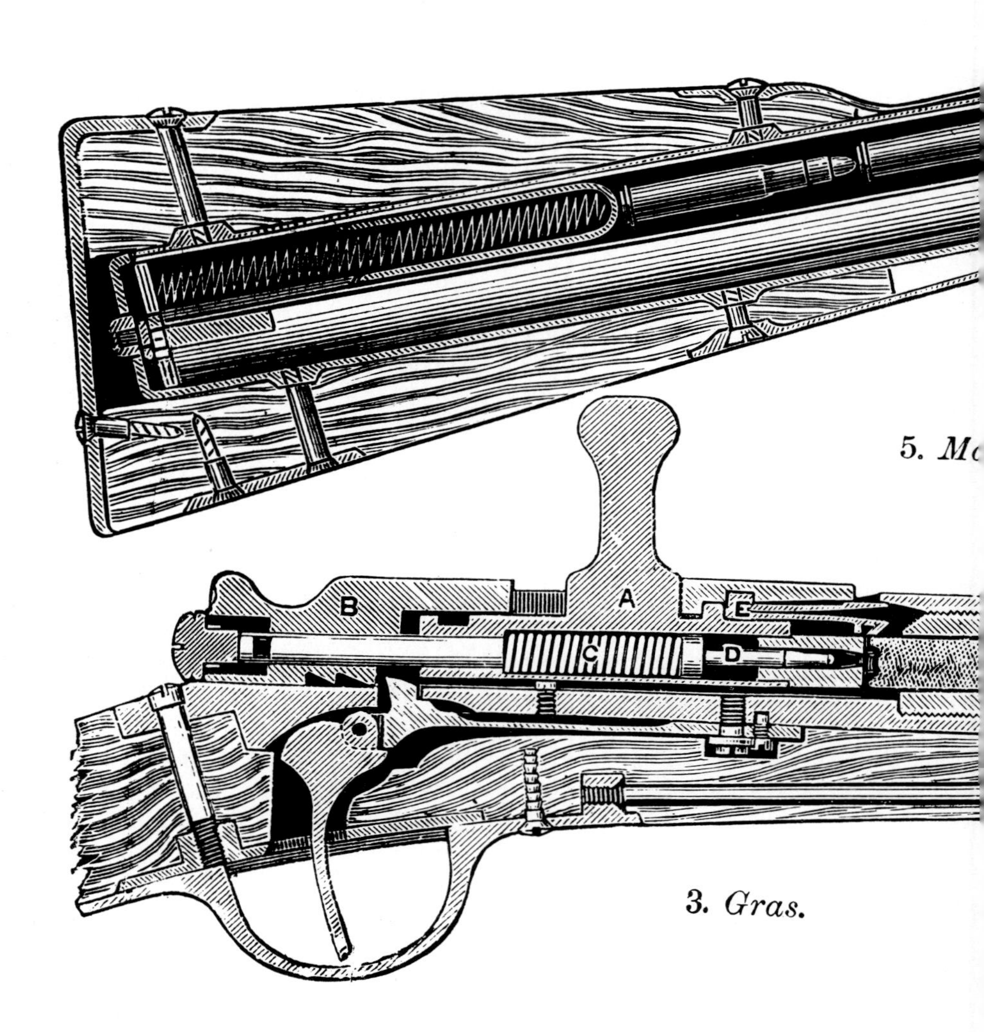

In the 19th century great attention was given in the technical and even lay press to comparisons of the latest military innovations; this diagram compares contemporary breech actions: the British Martini-Henry, the British Snider-Enfield, the French Gras, the German Mauser and the Austro-Hungarian Mannlicher.

SMALL ARMS.

DERN MILITARY BREECH ACTIONS.

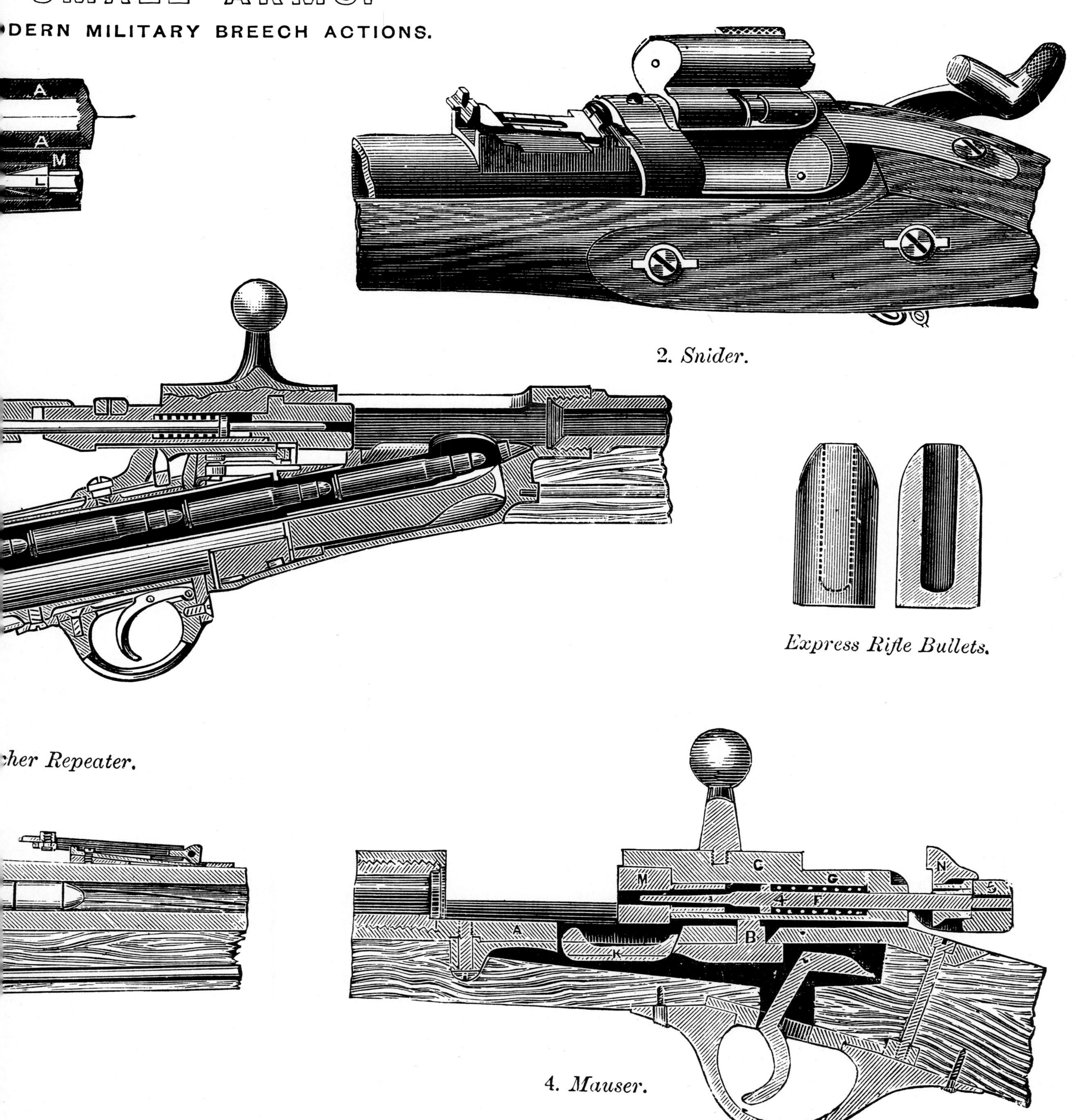

2. *Snider.*

Express Rifle Bullets.

cher Repeater.

4. *Mauser.*

plexity of armies, the carbine was also adopted as the primary arm of second-line forces, who might need a weapon in emergencies for self-protection and policing, but who would be hampered in their ordinary duties by the long military rifle designed to deliver accurate fire at ranges as great as 1000 yards/915 meters.

Pistols

The other main military weapon of the period was the pistol, designed for single-handed use. It was customarily issued to officers, who were deemed to need a hand free for signals and the like, and to military and civil policing units. From the earliest days of firearms development there had been attempts to make pistols, but it was appreciated that their short range left the firer little time for reloading and that a multiple-shot capability would thus be useful. So some of the earliest repeating weapon efforts were revolver pistols, in which a pre-loaded chamber fed a fresh round into line with the barrel as the previous round was fired.

A genuinely effective weapon of this type had to wait for the development of the fixed metallic cartridge, though progress thereafter was rapid. However, the first useful revolvers were the percussion-type weapons of Samuel Colt, who secured his initial patent for such a weapon in 1835. The Colt mechanism is simple: the cylindrical magazine is chambered for six rounds loaded (from the back once metallic cartridges were adopted), and a six-toothed ratchet cut into the rear of the cylinder is moved round 60 degrees each time the hammer is cocked, the cylinder being locked into accurate position (with a chamber exactly behind the barrel) by a spring mechanism. The early Colt revolvers used a six-chambered cylinder in which each of the chambers was loaded with a ball and powder (or with a paper cartridge) before the individual nipple for each chamber was fitted with its percussion cap. Considerable care was taken in the design of the cylinder to prevent the flash from any one chamber reaching and detonating the charges in adjoining chambers. The first such weapons were made in 1836, but a six-year gap followed when the company went bankrupt. By the mid-1840s the Colt revolvers were again available, and quantities increased rapidly as the virtues of the type were appreciated and as Colt firmly crushed the efforts of competitors to infringe his patents. The most important of the early Colts were the Navy Model 1851 and the Army Model 1860, but the basic concept received a new lease of life with the adoption of metallic cartridges, which greatly eased and speeded the loading process by means of a hinged-gate mechanism behind the chamber. Colt produced his first metallic-cartridge revolver in 1872, and the most celebrated of these weapons is the Model 1873, otherwise known as the Peacemaker and a feature of nearly every Western ever made.

Other manufacturers who got into the market once the Colt patents expired were Remington, Savage, and Smith & Wesson. The last contributed greatly to the type with the introduction of the hinged frame, in which the weapon is "broken" to give easy access to all the chambers. Initial models were broken forwards, but in 1892 the French devised a side-hingeing system in which the frame swung out to the left for reloading. A variation on the forward-breaking system was the Dodge fast-loading system, in which the breaking action operated a cam that lifted out all the cases in the cylinder to permit rapid reloading.

Far left, top:
Much of the British army's experience during the reign of Queen Victoria lay in colonial wars; this illustration shows a typical episode (the 144th Foot at Waitato Paa in New Zealand in 1863) where European firepower in the relative open proved too much for native forces.

Left, above:
The Battle of Inkerman in November 1854 during the Crimean War was a singularly bloody engagement. Both sides used rifled muzzle-loaders (the British the superior new Minié) to good effect as the large numbers on each side made continuous volley fire a practical proposition.

Left:
The Wagon Box Fight *by H. Charles McBarron illustrates the type of action sometimes faced by the US Army in the West. This engagement in Wyoming during 1867 involved the usual blend of single-shot breech-loading rifles and Colt revolvers.*

Below:
Mêlée engagements, such as depicted here in the Battle of Kandahar in 1880, called for rifle-equipped infantry supported by cavalry using carbines and, at close quarters, pistols. The latter weapon would have been much appreciated by the artillery being overrun by the 2nd Prince Edward's Own Gurkhas and the 92nd Highlanders.

Modern small arms are born

Though great improvements had been made in the reliability and consistency of gunpowder over the centuries it had two severe military limitations: it produced large quantities of dark smoke and its detonation was too quick for effective use in barreled weapons. Prolonged firing produced a thick fog which impaired breathing and visibility and revealed the firer's position. The second factor dictated in great measure the design of various weapons: high-quality grained gunpowder explodes (extremely rapidly in most cases) and so imposes enormous strains on the breech, the barrel and even the bullet because of the sudden and extremely high accelerations resulting from the virtually instantaneous detonation.

Propellants

It was the military liability of smoke clouds that led to the initial research into what were at first termed smokeless propellants, and though it was the Germans who led the way, it was the French who reached the goal first. In fact the development of smokeless propellants was more an evolutionary process than an invention. As early as the 1830s French chemists such as Braconnet and Pelouze had begun work on low-smoke propellants, but it was the German Schonbein who really showed the way with his guncotton in 1846, the German Hartig who showed how the rate of combustion could be controlled in the 1860s, and the Germans Schultz and Volkmann who developed the first primitive smokeless propellants in 1867 and 1871 respectively. In 1884 Vieille was able to cap all these efforts by producing what the French government was so desperately seeking to redress the military balance that had swung so heavily in favor of the Germans during and after the Franco-Prussian War. Vieille's propellant was essentially a mixture of nitrocellulose and picric acid, the latter soon replaced by ether and alcohol. Other nations were working along parallel lines, and the most successful of these in the late 1880s resulted in the British cordite, a mixture of nitroglycerin and guncotton turned into a gel by a solvent

Below:
The first machine-gun adopted by the US Army was the gas-operated Colt-Browning M1895, at first chambered for an indifferent 6 mm cartridge and later developed in the standard 0.3-inch caliber. The type was adopted by the US Navy in 1896, and was later upgraded into the Marlin machine-gun that found fame as an aircraft weapon.

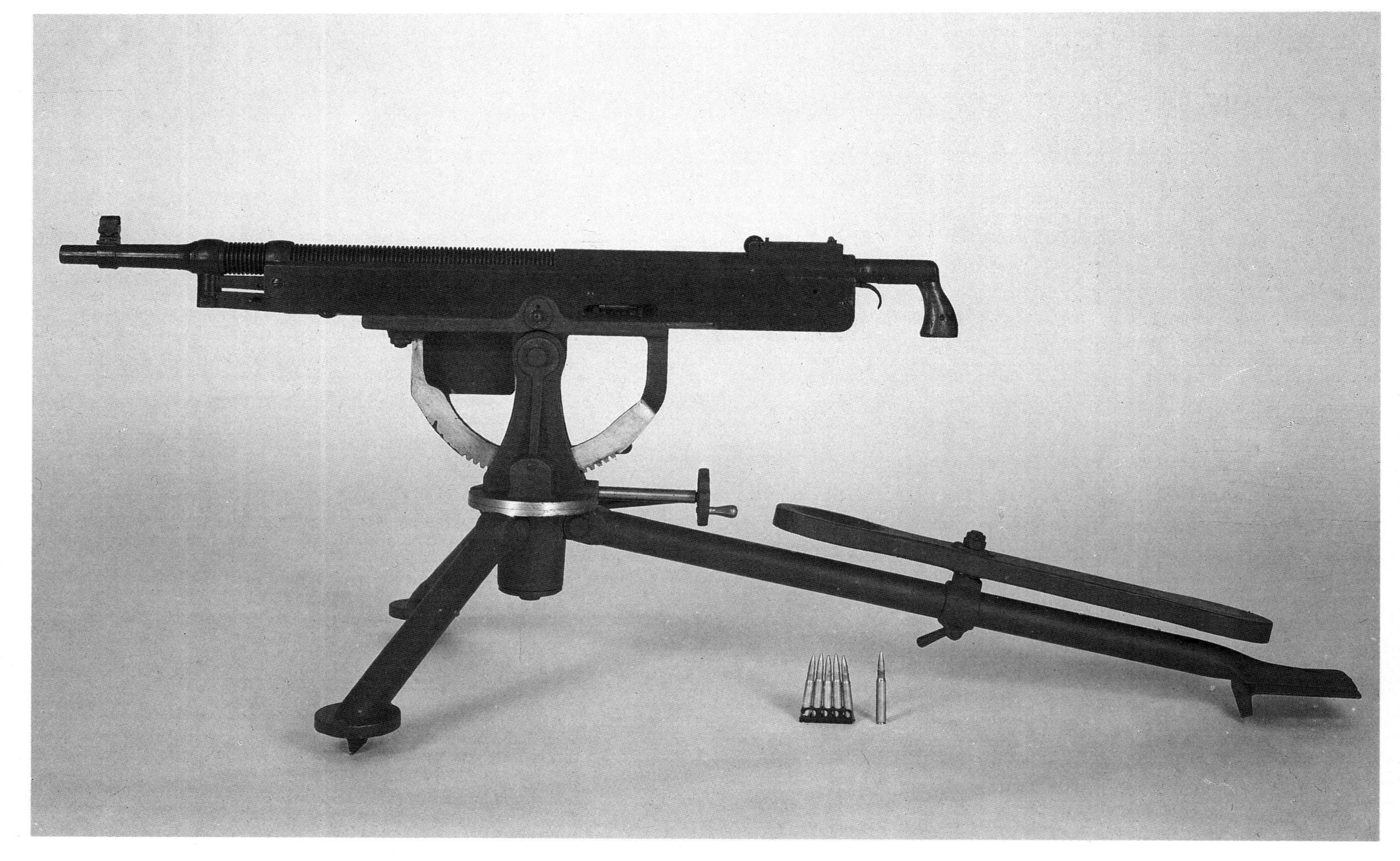

Left:
Later Gatling guns were notable for the excellence of their brass finish, and this example has the patent Accles Magazine that did much to keep foreign bodies out of the temperamental workings of the Gatling.

and then solidified with a mineral stabilizer.

Vieille's propellant was remarkably successful, and freedom from smoke was just one of its advantages over black powder, as experimentation soon revealed. The more important properties of the new type were greater chemical and ballistic efficiencies (providing higher but also more consistent power than gunpowder), and smoother burning rather than detonating qualities. This meant that the mass of metal needed to contain the detonation of gunpowder could be reduced for the nitrocellulose-propellant weapons, a tendency further increased by the smoother acceleration making possible the use of smaller bullets that left the weapon at far higher muzzle velocities than those fired with black powder. A bonus for the average soldier was that the new propellant fouled the barrel far less than black powder, easing the work of keeping his weapon clean even if the designer was faced with a different set of barrel-wear problems.

Another problem now facing the designer, however, was that the previous lead bullet was incapable of handling the accelerations (both radial and axial) imposed by the new propellant. The solution was already in train as a result of problems with the radial accelerations imposed by the improved rifling systems: the jacketed bullet with copper or cupro-nickel over a lead core. This could also take the stresses of the new propellant, and also had better ballistic qualities from the higher length-breadth ratio of the longer but slimmer bullet, and the inward movement of the center of mass.

The French rifle designed to use the new propellant was the Lebel Model 1886, which had a bolt action and an eight-round tubular magazine, but at 8 mm/0.315-inch a caliber markedly smaller than that of most black-powder contemporaries. The muzzle velocity was considerably greater, however, at 2379 feet/725 meters per second. Other nations were not far behind: Mauser designed the 7.65 mm/0.3-inch Model 1889 for Belgian manufacture before developing the Model 1893 for Spain and finally the definitive Gewehr 1898 for Germany in 7.92 mm/0.312-inch caliber. Austro-Hungary used a number of cleverly designed Mannlicher weapons. France adopted the Berthier carbine, Italy had Carcano-designed but Mannlicher-inspired rifles, and the UK developed the Lee-Metford into the Lee-Enfield, perhaps with the Gewehr 1898 the definitive bolt-action rifle of all time.

The effect of these changes was radical, for the theoretical range of black-powder weapons became the practical range of nitrocellulose weapons, opening the ranges of tactical engagement on a battlefield and improving visibility for soldiers who could carry far larger numbers of rounds for weapons that were easier to handle and lighter to fire.

Top:
Hard Pushed, *a painting by Charles Schreyvogel, typifies the type of role the 0.45-inch Colt Model 1873 revolver is best remembered for, and where it indeed performed to notable advantage.*

Right:
Though this American illustration depicts mainly the khaki uniform of the US Army between 1898 and 1900, the standard Krag-Jorgensen bolt-action rifle is also fully in evidence. This model was obsolescent by the turn of the century, and was replaced by the highly capable Springfield M1903 with a box magazine.

Right:
All That Was Left Of Them: *the last stand of the 17th Lancers in a Boer War engagement reveals the service carbine of the day, the Lee-Metford.*

Below:
The Battle of Abu Klea during the Sudan War of 1885 was another indication that the breech-loading rifle was highly effective; however, problems with the coiled brass cartridges could almost cause its operators to be crushed by superior manpower.

Automation

The advent of the magazine combined with the new propellant to open another major avenue of development to weapon designers and hence the armies of the world. This was the automation of the ejecting/cocking/loading cycle, long a dream of designers and operators alike as a means of raising the rate of fire by leaving the firer free to concentrate on aiming. Clearly such a system was only feasible with a magazine of ammunition on which the automatic portion of the operating cycle could draw, and the clean-burning properties of nitrocellulose propellants made it possible to use the surplus gas energy in the barrel as the dynamic force for the operation.

Properly speaking, the type of action desired by the designers was semi-automatic, or self-loading after the manual loading of the first round. There are several ways in which this objective can be attained, but safety considerations dictate that full-power weapons such as a rifle be fired only from a locked breech, so the gas-operated system was and is virtually universal for self-loading weapons.

The basics of such a system are simple: as the first round is fired after manual loading, a very small portion of the gases driving the bullet up the barrel is tapped off through a small hole in the barrel, turned through 90 degrees and used to drive a piston (though some systems do without the piston). This rearward motion unlocks the breech and drives back the bolt (in the process extracting and ejecting the spent case) against the pressure of a return spring, which takes over after the energy of the gas/piston system has been expended to push the bolt forward (in the process stripping a fresh round from the magazine and chambering it), cock the action and lock the system after it is completely closed. At this point the soldier can pull the trigger to fire a round and initiate the reloading cycle once again. Quite apart from the advantages of greater rate of fire, a semi-automatic does not require the soldier to remove a hand from the weapon to operate the bolt action, unsteadying the rifle and thereby spoiling the aim.

The first man to come to tackle the problems of semi-automatic action was an American, Hiram Maxim. After early work in which he used the recoil of a Winchester rifle to work a system of springs and lever that reloaded and recocked the weapon, in 1884 Maxim developed a recoil-operated semi-automatic rifle mechanism, but soon found that it was relatively unsafe and that the mechanism was too heavy and bulky for practical use in a rifle whose loaded weight should be no more than 10 lb/4.5 kg or so. Maxim turned his attention to a gas-operated system offering the possibility of reduced weight and bulk. The first successful Maxim gas-operated self-loading rifle was an 1891 conversion of the British Martini-Henry, but Maxim thereafter turned his attentions almost completely to fully-automatic guns (machine-guns), abandoning further development of semi-automatic weapons. Nevertheless, Maxim's work was widely appreciated, and came to form the basis of many other semi-automatic weapons.

At much the same time as Maxim, Mannlicher was

The Boers had to get modern weapons from any available source, and though the Mauser rifles were popular, many Lee-Metfords and Lee-Enfields were taken from the British. These types are evident in this nicely-posed group of burghers at Colesberg during February 1900, during the later stages of the Boer War. Personal kit may have been sketchy and worn, but the rifles and ammunition bandoliers are in good condition.

Above:
At the Battle of Tamai in the Sudan during 1884, the British used the Martini-Henry rifle to good effect despite suffering considerable problems with the coiled brass cartridge of the weapon.

Far right:
Looking highly impressive in his smart uniform and carrying a Snider rifle, a Sikh sentry stands guard at Fort Johnston, Nyasaland, during 1894.

working on a similar recoil-operated system in Austro-Hungary. Although the Mannlicher system was generally less refined than Maxim's, it did use perhaps the first accelerator, a pivoted device to speed the movement of the breech-block after the breech had been unlocked. Little came of Mannlicher's work in the short term, but it is interesting to note that John Browning was at the time working at the Fabrique Nationale factory at Herstal in Belgium, an establishment that prided itself on keeping fully up to date with small arms development all over Europe. The pivoted accelerator was later a key component of Browning automatic weapon designs. There were many designers at work on the same lines at this time, and all were hampered to greater or lesser degrees by the inconsistent firing of contemporary black-powder cartridges.

All this began to change with the French development of the Lebel cartridge and the universal adoption of nitrocellulose propellants. The first real fruit of the propellant revolution, so far as semi-automatic weapons were concerned, was a Mannlicher weapon based on the 1885 design but greatly improved in detail. Again the system was gas-operated, but additions were a Mannlicher box magazine so that the weapon could be loaded from above, a device to hold the bolt back when the magazine was empty (so facilitating rapid reloading) and a cocking handle. Yet again, this fully practical weapon found no official favor, but Mannlicher worked on regardless, producing two more designs (derived from turn-bolt and straight-pull repeating rifles) in 1893, two further weapons (with solid breeches and moving barrels) in 1894, a locked-breech gas-operated rifle in 1895, and yet another locked-breech gas-operated rifle in 1900. Though none of these weapons was accepted for production, they were rife with innovative features: several items from the 1895 weapon were later used in the US Garand rifle, and the 1900 weapon contributed to the Lewis light machine-gun.

Some indication of the enthusiasm with which designers were trying to evolve satisfactory semi-automatic rifles is provided by just a partial list of experimental models: in Denmark the Bang rifle was produced in 1911, in France a

series of development models was produced at the St Étienne arsenal from 1894 onwards, in Germany Mauser developed an experimental but not very effective rifle in 1898 and two more workmanlike models in 1902, in Italy Cei-Rigotti produced a thoroughly workable weapon in 1900, in Japan Nambu produced a useful weapon in 1904, in Sweden the Kjellman rifle appeared in 1904, and in the USA Winchester and Remington developed self-loading rifles in 1903 and 1906 respectively. Some of these weapons contributed to further developments (the French and German models to a considerable degree), but the first semi-automatic rifle to be used in combat was, oddly enough, a Mexican design, the Mondragon. This was produced in Switzerland in small numbers in 1913 and 1914, and featured a 10-round box magazine and a simple yet effective gas-operated system: a few Mondragons found their way into the hands of observers in aircraft during the early months of World War I.

Machine-guns

Allied to the self-loading or semi-automatic rifle in many of the basic operating principles is the machine-gun, a fully-automatic weapon designed to produce large volumes of rapid and concentrated fire. The notion of such a weapon is virtually as old as firearms themselves, but the same practical constraints that prevented the development of an effective semi-automatic rifle until the later 1880s also applied to the machine-gun. This is not to deny, however, that ingenious designers had tried, sometimes with a measure of success, to develop weapons of this basic type as volley guns (with many barrels and designed to deliver a simultaneous volley, for example a few 15th century organ guns and the 25-barrel Billinghurst Requa volley gun of the American Civil War) and crank guns (using a hand crank to operate multiple chambers for a single barrel, such as the Ager Coffee Mill Gun of 1862).

USA

The American Civil War in fact proved a decisive forcing ground for the machine-gun, as it did for several other modern weapon types, and two of the more interesting results were the Williams and Gatling hand-operated machine-guns. The former was designed to fire 1 lb/0.45 kg rounds with a caliber of 1.57 inches/39.88 mm: one man turned the crank that operated the sliding breech-block and another fed in paper cartridge rounds, and it could fire an estimated 65 rounds per minute. The Williams gun was so heavy that it had to be moved on an artillery carriage, but the Gatling was an altogether handier design, at least from the tactical point of view, and may be judged the world's first successful (albeit mechanical rather than gas-operated) machine-gun: a hand crank turned the mechanism that loaded, cocked, fired, extracted, ejected and reloaded the weapon, which had six 0.57-inch/14.48 mm barrels turning round a common center in its earlier forms. The whole device was practical if somewhat heavy and cumbersome, but was designed for protecting important positions and lines of communications rather than for field operations. Further development improved the mechanical reliability of the Gatling gun, so that at times they were deployed in the field, and until the advent of more advanced weapons the Gatling type of machine-gun was not uncommon in the USA and Europe.

Right:
The weight of the Gatling gun and its ammunition made it a semi-mobile weapon best treated as an item of light artillery, used for movement of a small carriage with attached limber for tools, ammunition and perhaps a section of the gun crew.

Below:
The 1st Punjab Infantry (Coke's Rifles) show off their Lee-Enfield rifles during 1902.

MOUNT
1942

One of the truly great weapons of all time, the 0.303-inch Vickers machine-gun was thoroughly reliable and, given proper maintenance and handling, capable of prodigious performance in the sustained-fire role.

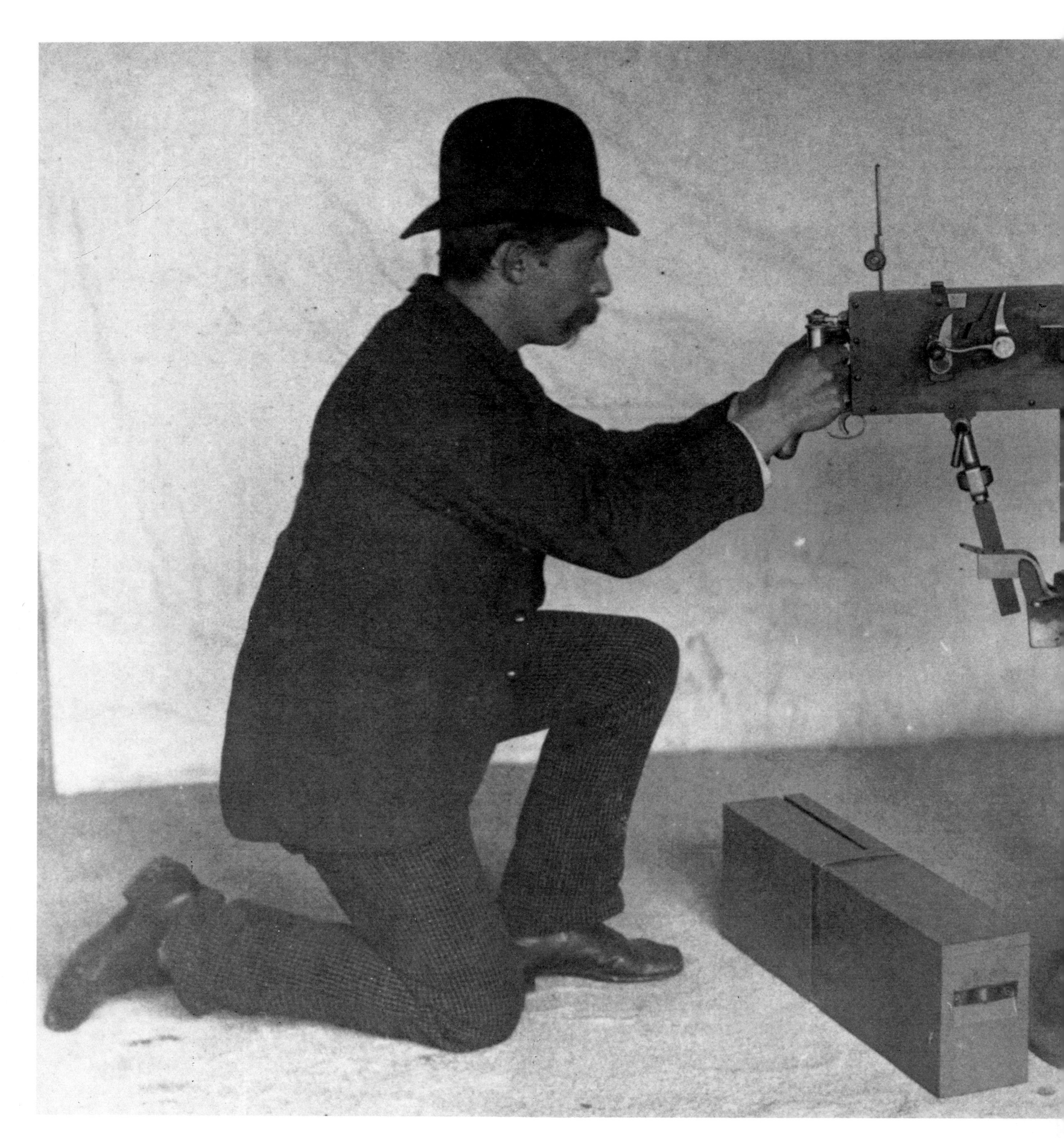

FRANCE

The French placed great faith in a comparable weapon during the Franco-Prussian War, but were sadly disappointed in the deReffye Mitrailleuse, which was in essence a frame containing 37 (later reduced to 25) rifle barrels. The loading device was a metal plate with ammunition holes corresponding with the rear of the barrels: after the breech-block had been moved back, the ammunition-loaded plate was dropped into slots at the sides of the breech-block after which the breech-block was screwed forward to drive the rounds into the barrels, where they were fired by rotating a handle. With a ready supply of ammunition plates and an experienced crew, the French reckoned on a rate of fire as high as 300 rounds per minute. The whole device was cumbersome, weighing over 2 tonnes, and the French considered it almost as a piece of artillery to be used en masse where they were quickly picked off by the Prussian artillery, whose light field pieces far outranged the Mitrailleuses.

The great name in the early days of French machine-gun design is Benjamin Hotchkiss, a prolific innovator who died in 1885 but left a thriving small arms company, which in 1893 was offered the rights to a gas-operated machine-gun designed by Baron Adolf von Odkolek, an Austrian. Odkolek had chosen his moment well, for the earlier Hotchkiss designs (including various Gatling type weapons) were faring badly in the world markets against the Maxim gun. Though Odkolek's weapon was not particularly effective in its rudimentary form it was air-cooled (and thus lighter than the Maxim) and it introduced a refinement to the standard gas-operation principle in that the gas piston was in a tube under the barrel (so offering strength and compact lines). The shrewd directors of Hotchkiss refused a license agreement, and instead bought the rights to the Odkolek design, which by 1897 had been refined into a moderately useful locked-breech weapon operated by gas tapped off the underside of the barrel and using the standard 8 mm Lebel round fed from an unusual metal strip system (24 or 30 rounds) on the left of the gun. The French of the period were inveterate tinkerers, and variations on the original Hotchkiss theme were produced in the form of the Puteaux, St Étienne and Benet-Mercie machine-guns, all unsuccessful largely because they changed a workable design for no good reason. The definitive Hotchkiss was the Model 1914, which consolidated the improvements made in preceding weapons to produce a truly effective lightweight machine-gun.

UK

As with the self-loading rifle, the machine-gun owes its practical origins to Hiram Maxim, an extremely talented engineer who had made his fortune in the USA with a host of far-sighted schemes before moving to the UK. Maxim had already experimented with self-loading rifles using recoil energy for the reloading cycle, but rapidly saw that the principle was more suited to a fully automatic weapon such as a machine-gun, whose lack was sorely felt by most European armies. Maxim thus set to work and in 1884 produced his first machine-gun, which worked perfectly from the start of firing trials as a result of the design expertise of Maxim and the skill of the machine operators who had faithfully implemented his desires using the latest

A Maxim gun on a pedestal mounting shows how the type could be used on board ship, with a neat shelf for the ammunition box.

Left:
The storming of the Taku Forts (seen here as German sailors cross the outer defenses) during the Boxer Rebellion confirmed that disciplined soldiers, armed with modern weapons such as the Gewehr 98, could face and defeat far larger forces using less capable weapons.

Right:
Using their Gewehr 98 rifles to remarkable effect, German pioneers storm a defended pass during the Boxer Rebellion of 1900.

Below:
The Boxer Rebellion of 1900 in northern China was a near-disaster for the European powers in China, and only concerted action by a multi-national force saved the day. Here German sailors armed with Gewehr 98 rifles engage Boxers trying to prevent the relief force from reaching the beleaguered legations in Peking.

machine tools. The system used by Maxim was short recoil: as the first round was fired, the barrel and bolt were locked together, recoiling for about 0.75 inch/2 cm before the barrel was halted and a toggle mechanism unlocked the bolt. The bolt continued to move back, compressing a return spring which then shot the bolt forward again to strip a fresh round from the feed mechanism, chamber it, lock itself to the barrel and then release the firing pin to discharge the next round and start a new cycle. The feed mechanism was a canvas belt holding 333 rounds, the ends of which were linked together for continuous fire at variable rates controlled by a variable oil buffer device set by a lever on the right of the weapon's receiver.

Despite using black powder rounds, this 1884 model worked reliably and effectively. The British army was quick to recognize the significance of the new weapon, and ordered a production model from Vickers, which began to enter service in 1891. Its efficiency was proved in the Matabele War of 1893-94 when a group of 50 British infantrymen, armed with rifles and four Maxim guns, were faced by some 5000 Matabele. In a 90-minute engagement the Matabele charged the British position four times, and each time were repulsed by the well-sited Maxims, losing 3000 dead in all. Similar successes became frequent in the border and tribal wars of the British Empire, and though some maintain that they were over indifferently led and spear-armed tribesmen such as the Matabele, in fact most victories were over the wily and highly capable hill tribesmen of the North-West Frontier regions of India. Even so,

these were police rather than military actions, and it was the success of the Maxim gun in the Battle of Omdurman during the Dervish War that secured its total acceptance: the British commander General Sir Garnet Wolseley had been an advocate of the Maxim since its earliest days, and made sure that some were available for his troops in this Sudan war. Of the 20,000 or so Dervish dead at Omdurman, about 15,000 are thought to have been killed by the Maxim guns of the British force, which suffered the extraordinarily low casualty rate of only two per cent against vastly determined opposition.

The tactical limitation of the machine-gun was weight, with the result that the early operators used it only for covering fire in an attack, and for defensive fire as part of the defenses of a fortification. In this policy they were supported by the weight of the water-cooled Maxim and its heavy mountings, though it was soon apparent that the wheeled carriage could be left to attract the attention of the enemy while the gun itself was moved to a tactically more advantageous position. But in the Russo-Japanese War of 1904-5 the machine-gun began to come into its own as a mobile attack weapon: the Russians used the 7.62 mm Maxim initially on a high-wheeled mounting with a shield, and the Japanese used the lighter French-designed 8 mm Hotchkiss. On several occasions machine-guns proved decisive, but the main lessons of the campaign emphasized the defensive capability of the Maxim, especially when used on a low mount, and the relative unreliability of the air-cooled Hotchkiss.

GERMANY

Oddly enough, the Germans at first saw little need for the machine-gun, and it was only in 1899 that the German army adopted the Maxim and launched an intensive development programme that produced the excellent MG08 in the standard German caliber of 7.92 mm. This model was produced at the state arsenal of Spandau, a name often applied generically to the MG08, which was a fairly heavy

Indian cavalry show off their Snider carbines in the last decade of the 19th century. Though single-shot weapons, these were light, handy, and well-suited to the perceived need of the cavalry of the period, who used the carbine during the earlier phases of an engagement and then closed with the sabre.

weapon fed from 250-round fabric belts. The other major German machine-gun type of the early period was the Bergmann, whose origins lay with a prototype of 1900. The definitive model appeared in 1902, and though the German army preferred the Maxim, the Bergmann was produced as a private venture and proved itself an excellent weapon in its own right. The Bergmann used the short-recoil system and was water-cooled, but its real advantages lay in three special features: the quick-change barrel, the ammunition belt and the feed system. The ability to change the barrel without losing water was very useful, as other water-cooled weapons overheated despite their coolant and had to be pulled out of action. The ammunition belt was made of non-disintegrating metal links and was specially designed for use with the feed system, which was capable of accepting (and repositioning) badly aligned rounds that would have jammed any contemporary cloth-belt weapon. Another German machine-gun with good feed characteristics was the Dreyse, designed by Louis Schmeisser but named in honour of Johann Dreyse in 1907. The weapon was produced in small numbers only, but set the manufacturing company on the path of automatic weapon design to the great advantage of Germany in World War II. This concern was the Rheinische Metallwaren und Maschinenfabrik AG, later known as Rheinmetall. The other major German machine-gun of the period (and of World War I) was the Parabellum, produced in 1911 by DWM as a radically lightened version of the Maxim (MG08) with air-cooling for better tactical flexibility.

AUSTRO-HUNGARY

Other European countries were soon in the field with their own machine-guns. Austro-Hungary were highly impressed with the Maxim, but rather than adopt a license production system opted for a far poorer indigenously designed weapon. This was the 8 mm Schwarzlose, in fact designed by a German in 1902 and launched into Austrian production during 1905. Until modern times delayed-blow-

Above:
The Sharps rifle was phenomenally accurate for its time, and acquired a mystique that still lingers with the design. The Still Hunt *of 1888 by J.H. Moser shows a buffalo hunter in action with his powerful Sharps weapon.*

Left:
Armed with Martini-Henry rifles, men of the 5th Gurkhas and 72nd Highlanders storm Pewar Khotal in 1878 during the Second Afghan War.

Right:
The Zulu War of 1879 was a British victory—but it showed that courageous masses could virtually overwhelm smaller but well-equipped European forces, if the attackers were prepared to accept heavy losses.

back action for machine-guns has been rare, and indeed the Schwarzlose was the only successful mass-production machine-gun of this type until after 1950. Schwarzlose's adoption of the delayed-blowback system was on economic grounds, for the very close tolerances of recoil- and gas-operated machine-guns required extensive and costly machining. Schwarzlose opted for a strongly made but comparatively simple delayed blowback action which was less safe as it fired from an unlocked breech but was much cheaper to make.

Recoil operation is based on the premise that when a round is fired in an unlocked breech, as the bullet starts to move up the barrel the unlocked breech begins to open under the influence of the recoil forces, but can be controlled by powerful springs and a mechanical advantage system (without need for a massive breech-block). This means that even with a full-power military round the bullet has left the necessarily short barrel before the breech has opened significantly. Thereafter the operation is conventional, the breech recoiling to its maximum extent before being returned to battery and in the process stripping and chambering the next round. There were considerable teething problems with the type, which in its earlier forms required the individual lubrication of each round by an action-operated oil pump, but by 1912 the Schwarzlose reached a stage of adequate development. But while it was considerably cheaper than comparable gas- and recoil-operated weapons, it was still a heavy water-cooled weapon fed from a fabric belt.

DENMARK

A similar recoil operation is used by the Danish Madsen machine-gun, which was introduced in 1902 after development from the mid-1880s. The type is notable for its longevity, and very many years of successful operations have put the lie to claims that it is unsafe, which is based on the Madsen's unique variation on the recoil theme: the action is a modified Peabody-Martini one, the opened breech-block being an oscillating mass that drops below the bore to strip a fresh round and chamber it before rising again into battery. Thus the Madsen has the only non-ramming action in machine-gun design. Another advantage is its lightness as it is an air-cooled weapon.

3

ITALY

Though now little known for its machine-gun designs, Italy in the early period of automatic weapon development produced one of the truly outstanding designs of all time, the Perino of 1901. The Italians were highly impressed with this weapon, but decided to keep it secret, buying large numbers of Maxims to cover the fact that the Perino was to be built in conditions of great security. The best feature of this innovative water- or air-cooled machine-gun was the feed system using trays (each holding 25 rounds) fed in individually from a box on the left of the receiver. This insured consistent feed as the rounds were well aligned, and any jam was quickly cleared by pushing a button that ejected the troublesome round. The weapon had many other outstanding features, but Italy's delay in producing it meant that it was too late by the time the nation needed it. It made do instead with the Maxim and the 6.5 mm Revelli, an indifferent air-cooled weapon with a combined short-recoil and blowback action firing from a bolt that could be described as locked only in the loosest sense of the word.

There were other machine-guns, of course, but these were the more important to appear in the period leading up to World War I, the titanic struggle that was to demonstrate in the most ghastly way the total dominance of the machine-gun in the static warfare it helped to establish.

Semi-automatic pistols

The development of nitrocellulose propellants and effective self-loading mechanisms also paved the way for practical semi-automatic pistols. Such weapons had been proposed as far back as 1664 by a commission in England, but the first effective one to enter production was designed by the American Hugo Borchardt, but made in Europe by Ludwig Lowe of Berlin in 7.63 mm caliber. (The round was designed by Borchardt himself, and was the precursor of the celebrated Mauser 7.63 mm round.) Borchardt had worked for Winchester, and his passion for self-loading weapons finally resulted in this history-making weapon using a cumbersome but nonetheless workable toggle-lock. The weapon appeared in 1893, and in common with many other weapons of the type was marketed with a removable rifle stock as a means of insuring maximum accuracy.

Other weapons followed the Borchardt quite rapidly, typical examples including the Bergmann weapon of 1894, the Mauser military model of 1895 (the world's first self-loading pistol designed for military use), the Maxim of 1896, the Schwarzlose of 1898 and the Mannlicher of 1901. None of these early self-loading pistols was particularly successful, though they are of historical importance and interest, but perhaps more noteworthy as the weapons which introduced the cartridges that became standard in later and more effective self-loading pistols. The Bergmann round, for example, was developed into the 0.32-inch/8.13 mm Colt Automatic Pistol round and the 7.65 mm/0.301-inch Browning Automatic Pistol cartridge.

By far the best of these early weapons was the Browning Model 1898, designed by the legendary John M. Browning

Another effort to improve the tactical flexibility of the carriage-mounted Maxim was a light shield (similar to those used on small artillery pieces) to allow the type to be placed farther forward than usual.

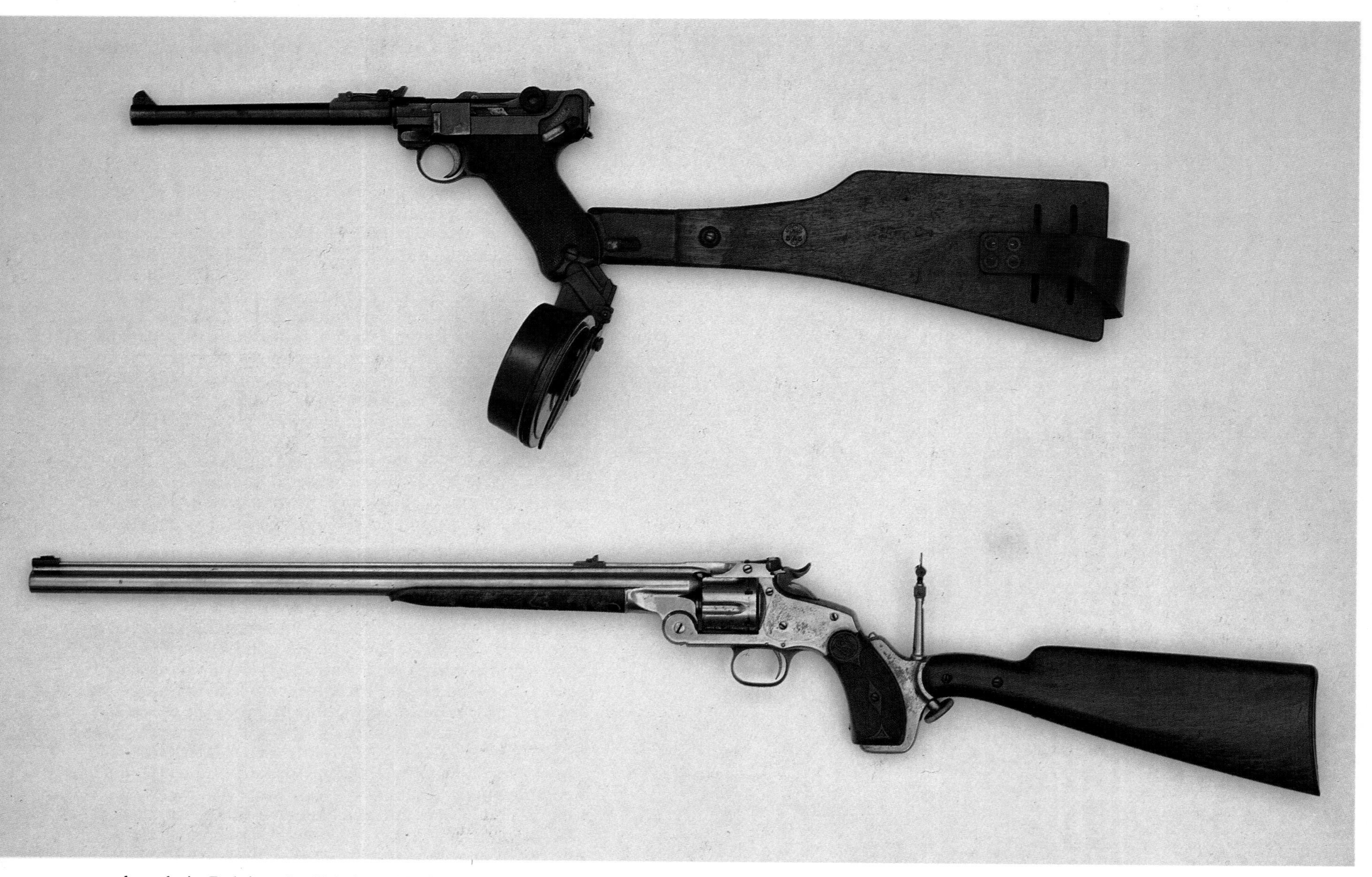

and made in Belgium by Fabrique Nationale as no American manufacturer was interested in this precursor of one of the best and most successful families of semi-automatic pistol ever developed. In the short term the most important of the series was the 0.45-inch/11.43 mm Colt M1911 designed by Browning for the Colt Company, and in 1911 adopted as the standard US service semi-automatic pistol, which it remains to this day (though a successor was chosen in 1985 for production in the near future).

Other notable weapons are the German Pistole 08 (far better known as the Luger after its designer) and the German Pistole 10 (far better known as the "Broomhandle Mauser" for its shape and designer). Georg Luger was an employee of Ludwig Lowe when the Borchardt pistol was being refined into production form, and played an important part in the streamlining of this clumsy American weapon. Luger saw, however, that a total redesign would produce a far superior weapon, and set to work on his own version, which was eventually produced in 1900 by Deutsche Waffen und Munitions-Fabriken (DWM) in 7.65 mm caliber. The type was adopted in 1908 as the standard German weapon, though the caliber was upped to 9 mm/0.35-inch in the new Parabellum round; the caliber and round have since become standard for self-loading pistols and sub-machineguns, but the Pistole 08 was their first application. The 9 mm Parabellum round was developed from the 7.65 mm type, so the caliber was simply changed by using a different barrel with bore and chamber appropriate to the larger caliber. Two of the main features of the Luger have always been its "pointability", leading to good natural aim as the gun becomes a simple extension of the firer's arm and hand, and an excellent combination of strength and reliability. The pistol is loaded by inserting a loaded eight-round inline magazine through the bottom of the butt until the magazine locks into place. The firer's free hand then grasps the milled knobs on the sides of the toggle mechanism, which is pulled up and back as far as it will go before being released to move forward under the power of the compressed recoil spring. This strips a round from the top of the magazine and chambers it, and the action then locks as the central hinge line of the toggle is below the other two hinge lines. This allows the connection of the trigger sear, and the weapon can be fired. As the pistol is fired, the locked-together barrel and recoiling mechanism drive back together for a short distance (allowing gas pressure in the barrel to drop to safe levels) before the central hinge of the toggle strikes an angled surface on the frame and is deflected upwards to unlock the weapon, and so initiate the extraction, ejection, cocking, reloading and relocking cycle again. The Luger can also be fitted with a so-called 'snail' magazine, in which a rotary extension to the inline box provides a magazine capacity of 32 rounds.

Above:
Early repeating and self-loading weapons were sometimes extremely ingenious in their basic concept, a good example being the Smith & Wesson Revolving Rifle (bottom), in essence a revolver pistol with a long barrel and detachable stock. This 0.32-inch caliber six-shot weapon was produced between 1879 and 1887. The other weapon is the Luger Artillery Model 08 9 mm Parabellum semi-automatic pistol (top), seen here with 32-round "snail" magazine and detachable stock.

Top right:
Two classic semi-automatic pistols of the early generation were the Borchardt (bottom), the world's first effective self-loading pistol, and the Mauser Model 1912, affectionately known as the "broomhandle

Mauser". Both weapons had long barrels, and were available with detachable stocks in an effort to capitalize on the accuracy possible with these lengths.

Right:
The Jameson Raid of 1896 sparked off the Boer War, which brought to prominence the advantages of bolt-action magazine rifles used by skilled marksmen. The linear tactics used by the British at the beginning of the war played right into the hands of the Boers, who knew the land and used camouflage and excellent musketry to decimate their opponents.

The Luger is a recoil-operated weapon, the Mauser a blowback weapon. The Luger may have the advantages of better "pointability", but the Mauser looks very purposeful, which indeed it is. The "broomhandle" is comfortable to hold, even if angled too straight down, and the magazine housing in front of the trigger group provides a good handhold for the free hand, so providing the opportunity for two-handed operation for greater steadiness and accuracy. Like the Luger, the Mauser can be used with a shoulder stock, a cleverly designed item of kit that doubles as a holster and also contains the weapon's cleaning and maintenance equipment.

These two German weapons are lightweights in comparison with the mighty Colt M1911, which was so successful from the beginning that there was only one development, the M1911A1 of 1926, which improved various external features without altering the mechanism of this potent seven-shot recoil-operated weapon. The magazine is in the butt, and locks audibly as it is inserted. At this point the firer seizes the roughened sides of the slide's rear portion and pulls the slide backwards against the pressure of the recoil spring, in the process cocking the hammer which is held by a sear but disconnected from the trigger for safety. The slide is now released, moving forward to strip a round and chamber it, allowing the trigger and hammer to reconnect. As the breech-block reaches the rear face of the barrel, the two move forward and up on the barrel link until fully forward, where they are locked to the inside of the slide by a series of interlocking grooves. The weapon can now be fired, the recoil driving back the locked barrel and slide until the safe point is reached: here the barrel disengages and drops, leaving the slide to move to the rear to repeat the standard cycle of extraction, ejection, cocking, reloading and relocking.

The period between 1885 and 1914 was thus a particularly important one in the history of small arms, and was made possible by the twin developments of fixed metallic cartridges and nitrocellulose propellants. These turned the previous breech-loading weapons into far more potent pieces with higher muzzle velocity and lower weight, and also made possible new weapons types because of the ammunition's more consistent qualities and greater strength. Earlier theories about semi-automatic and automatic weapons now became feasible, while parallel developments in metallurgy and machining made it possible to turn the designers' work into practical hardware, even if much of it was expensive. But it was a period in which international tensions were rising, and the affluence of many Western nations made it possible for them both to want and to procure the latest and the best.

Right:
Another early machine-gun was the Colt-Browning M1895, which was also installed on a light carriage to provide at least a measure of mobility. Note how near its center of gravity the weapon is mounted in comparison with other machine-guns of the period, which were heavier in the receiver portion of their workings.

Below:
The unwieldiness of the Maxim gun often meant that in less secure areas this powerful piece of equipment had to be protected by infantry or, as in this instance during the Boer War, by men of the Natal Carabineers.

The Vickers-Maxim Model 1899 was an effective machine-gun of the early type, chambered for the 0.303-inch round and already highly reliable. But the weight of the weapon and its cooling water meant that the type was best employed as a semi-mobile defensive gun mounted on a light carriage, as shown in this illustration.

1914 to 1939

Though there had been major conflicts such as the Crimean War, the American Civil War and the Franco-Prussian War during the Victorian age, most of these conflicts were before the most fruitful period of small arms development, so the lessons and tactical implications remained largely unexplored or confined to the largely irrelevant results of colonial wars. There were some lessons to be learned from the American Civil War and the Boer War, but for the European powers these conflicts were too remote to impinge directly on their armed consciousness. The same was true of the Russo-Japanese War of 1904–5, which was fought by modern armies with modern weapons, and which should thus have alerted all to the nature of the new warfare, which had been forcefully evolved from the conventions of most European armies by the effect of modern weapons, especially artillery and repeating small arms.

The start of World War I was thus marked by a military trauma that has left a deep scar on Western consciousness: the armies expected a short and tactically fluid war, only to be shocked emotionally and physically into tactical immobility. The weapons that produced this static and increasingly bitter novelty were rapid-firing rifles and machine-guns, backed by artillery and barbed wire. The whole nature of warfare had to be rethought as professional armies disappeared in a welter of lead and high explosive, to be replaced by armies that were in effect militia forces armed with vast quantities of the latest weapons the designers and factories could produce.

Below:
In the assault role the Bren could be fired from the hip to keep the heads of the enemy down, but accuracy was inevitably poor. To improve this, the weapon was used with the bipod folded, the right hand holding the pistol grip by the trigger and the left hand holding the barrel-change handle.

Left:
A Lewis machine-gunner ready for the start of an attack during March 1918, by which time the Lewis had come into its own as the Allied assault machine-gun par excellence.

Below:
The crew of a British armed trawler undertake a spot of arms drill under the watchful eye of a naval officer. The rifles are obsolete Lee-Enfields kept in reserve for just such eventualities.

Machine-guns

The quintessential weapon in this terrible transformation was the machine-gun. Light, medium and heavy artillery could wreak enormous havoc with the landscape and with major positions, and rifles could pick off large numbers of men moving over open ground, but when it came to the static operations that prevailed throughout World War I, except for the mobile opening and closing exchanges, the machine-gun undoubtedly ruled the battlefield.

In defense a small number of interlocking machine-gun positions could protect each other and pour a withering enfilading fire on massed infantry attacks moving across a churned no-man's-land and checked before the wire entanglements that were, with a ghastly inevitability, always imperfectly destroyed by the days'-long artillery bombardments that preceded set-piece attacks. The defensive machine-gun nests were also a primary target for the attackers' artillery. But at least some of the nests survived (perhaps because they were inadequately targeted or perhaps because they had remained undetected by ground or air reconnaissance) for their crews to pop up and man their weapons as the attackers moved out of their own trenches. They could then decimate the opposing infantry, or at least check them for long enough for the defense to bring up reinforcements.

In attack the machine-gun was less useful, though still a necessary adjunct to any major effort, because it lacked mobility as a result of its weight (weapon, mounting and cooling water could weigh in the order of 100 lb/45 kg) and the difficulty of moving up the prodigious quantities of belted ammunition it consumed in any type of protracted engagement. An immediate consequence of this limitation, which became apparent early in the course of the war, was the development of lighter weapons, generally air-cooled and designed to fire from lighter bipod mounts allowing the firer to lie down. Such weapons often used the belted ammunition of their heavier brethren, but many of the more advanced weapons were designed to use magazine feeds. These reduced the difficulties of transporting large quantities of ammunition and could also fire rifle ammunition secured from the accompanying infantry, and suffered fewer misfeed problems than the heavier weapons.

Heavy machine-guns

Most of these heavy machine-guns were already well established, having entered military service in the first decade of the 20th century or even the last decade of the 19th century.

FRANCE

The French used the Hotchkiss Model 1910 and Model 1914, the latter only marginally improved on the Model 1910. Both weapons were air-cooled (and might therefore be classified as medium machine-guns), but were used in the same way as the standard water-cooled heavy machine-guns, for long-range sustained fire. Apart from the unusual Hotchkiss strip feed discussed above (straight 24- or 30-

Three lesser known rifles: from top to bottom, the Ross Rifle Mk III, the Farquhar-Hill Rifle and the Pedersen T2E1. The Ross was a worthy design too delicate for extensive front-line deployment, the Farquhar-Hill was a 1908 British attempt at a recoil-operated self-loader with a 20-round magazine, and the Pedersen was an American attempt made in the 1920s to convert the Springfield to self-loading with a blowback action and 10-round magazine.

Left:
French Zouaves prepare to defend St Étienne airfield with an 8 mm St Étienne Model 1907 machine-gun, another variation on the standard Hotchkiss theme that failed to match the original in reliability and capability.

Below:
US machine-gunners of the 77th Infantry Division learn the intricacies of the Vickers machine-gun from a sergeant of the Machine-Gun Corps during May 1918. Accurate positioning of the round in the belt was essential to avoid jams: another potential problem was a wet belt, whose fabric could expand and get stuck in the feed mechanism.

round clips for the model 1910 and, in addition to these for the Model 1914, a 249-round semi-articulated belt made up of three-round links), the most notable feature of these weapons was an additional cooling device in the form of five 'doughnuts' round the barrel just forward of the receiver. Made of brass or steel, and making the rear of the barrel look as though it had a concertina collar, these doughnuts added to the mass of metal to absorb the heat of firing, and also provided considerably more area from which the absorbed heat could be radiated. Production of the Hotchkiss guns was stepped up enormously during the course of the war, for the design proved sound and the troops at the front made increasing demands for the Hotchkiss to replace the Hotchkiss variants produced by government arsenals, which as noted above had been changed merely for the sake of change and proved totally unreliable.

The demands on Hotchkiss were increased yet further by the arrival of American forces in France late in 1917 and during 1918. The Americans were the last modern nation to adopt automatic weapons wholeheartedly. The Americans in France thus lacked adequate numbers of indigenous machine-guns and called for Hotchkiss weapons, despite some three years of trench warfare having confirmed the

Hotchkiss machine-gun's inadequate feed reliability, especially in muddy conditions. The gun itself weighed 23.6 kg/52 lb and was usually fitted on a substantial tripod mount with a screw-elevation device for fixed-range fire; the rate of fire was between 400 and 600 rounds per minute using the standard Lebel 8 mm/0.315-inch ammunition.

UK

The British heavy machine-gun was the Vickers Mk I, a classic of its type and a weapon still encountered in less advanced corners of the world. The Vickers was essentially the production version of the Maxim for the British army, though significantly lighter than the Maxim: the Vickers engineers restressed the weapon and discovered that the Maxim was unnecessarily massive in certain features. Vickers also inverted the Maxim operating system, with the toggle lock opening up rather than down, making it possible to lighten the lock considerably. The operating system was in itself simple but extremely strong, and of the recoil type: the centrally-hinged two-arm toggle was in line and thus locked at the moment of firing; however, a muzzle trap caught some of the gas which was used to help drive back the combined barrel and breech-block to a point at which the rear arm of the toggle hit an angled post and broke upwards, unlocking the mechanism to complete the cycle of extracting, ejecting, cocking and reloading.

The Vickers Mk I weighed 40 lb/18 kg without water, and was usually mounted on a tripod weighing 48.5 lb/22 kg. Like the Hotchkiss gun, the Vickers tripod had a fixed-elevation screw, and it could take a bubble sight for night and indirect fire. Fed with 0.303-inch/7.7 mm ammunition from a 250-round fabric belt, the Vickers could fire virtually indefinitely at between 450 and 500 rounds per minute as long as the water coolant was kept topped up. This sustained-fire ability was realized early in the war and proved immensely valuable, though at first the plume of steam from boiling coolant was a tactical giveaway: the jacket held 7 pints/4 liters of water, which boiled after just three minutes of sustained fire at 200 rounds per minute. A condenser in the pipe leading the water back to the reservoir solved the problem. At the beginning of the war two Vickers were issued to every infantry battalion, but their importance grew so rapidly that the Machine-Gun Corps was raised to operate the type, it being realized that a dedicated formation well versed in the intricacies of rapid jam-clearing was far superior to increased allocations for standard infantry battalions. Another skill particular to the Machine-Gun Corps was rapid barrel changing: even with

Below:
Indian machine-gunners man a Benet-Mercié Model 1909 light machine-gun, derived from the heavier Hotchkiss weapon but notably unreliable in operation. The type's main failings were the inverted position of the feed strip (with the cartridges underneath the clip) and a poor locking system.

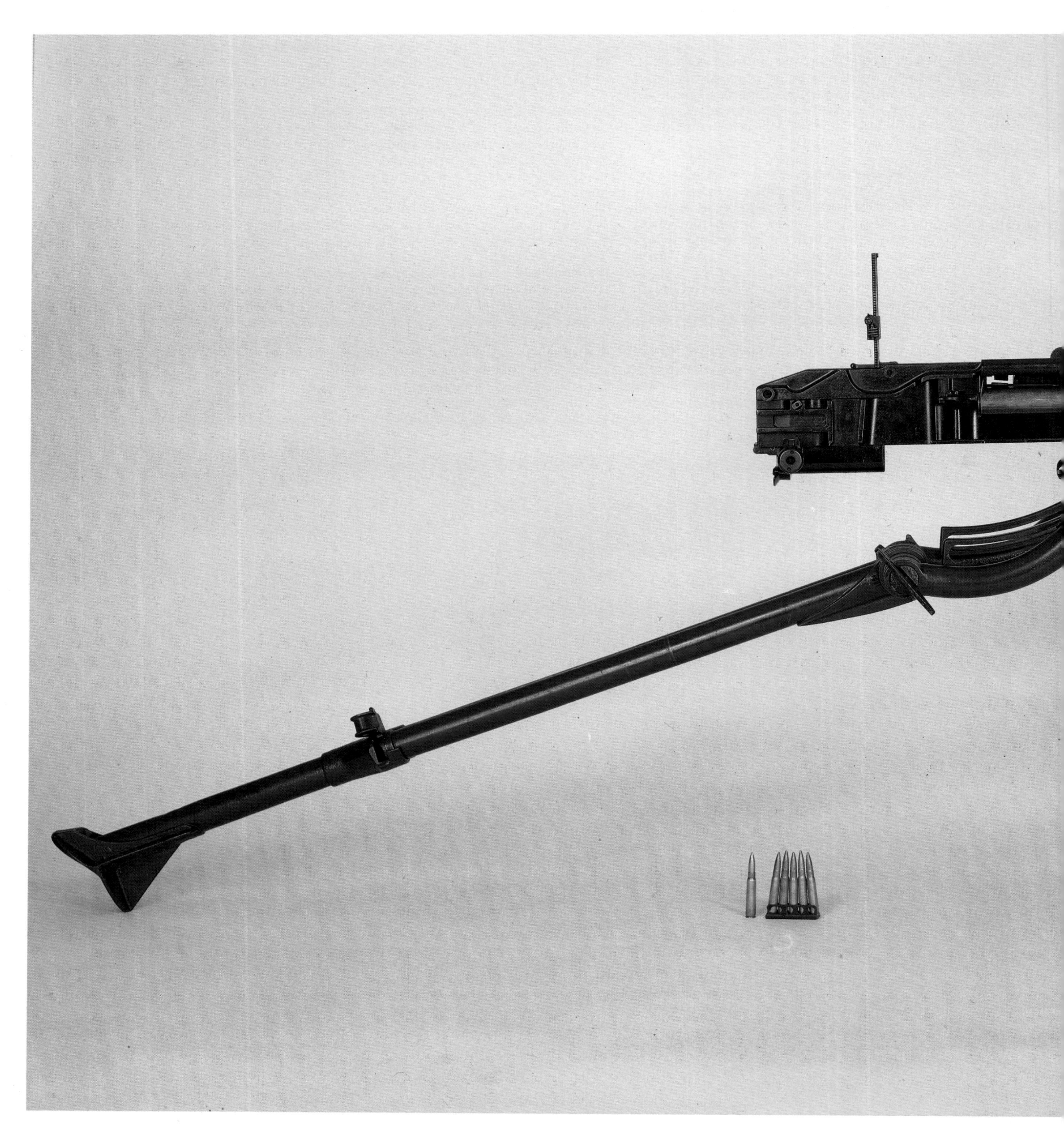

an uninterrupted supply of coolant water the barrel had to be changed every 10,000 rounds, and as a hectic action could consume this number of rounds in less than an hour, a rapid barrel change was tactically vital. A skilled crew could manage the task in as little as two minutes without appreciable loss of water.

Given its own operating formation, experienced crews and supply personnel, and the pressing tactical requirements of set-piece attacks in static warfare, it is hardly surprising that the Vickers should come to be regarded almost as a piece of light artillery. An illustration of this point, which also highlights the role of the heavy machine-gun in World War I, is the operation by the 100th Machine-Gun Company in the battle for High Wood (within the Battle of the Somme) in 1916. On August 24 it was decided that an infantry attack would be supported by the company's 10 Vickers machine-guns under cover in the Savoy Trench. It took two companies of infantry to stockpile the necessary ammunition, and during the attack the 100th MG Company fired continuously for 12 hours: carefully sited and zeroed in, the guns had a barrel change every hour, and the gunners and loaders were changed frequently as a deluge of fire was kept up on the predetermined targets designed to support the initial attack and to prevent any German counter-attack. The whole operation worked like clockwork, and in the course of 12 hours the 10 Vickers fired nearly a million rounds.

RUSSIA

Russia used its own version of the Maxim, the Pulemet Maksima 1910 which was identical with the PM05 initial model apart from a steel rather than bronze water jacket. Massive and expensive, the PM1910 was nonetheless a magnificent weapon ideally suited to Russian (and later Soviet) requirements of availability and utter reliability as attested by continued production right up to 1943, giving the PM10 the longest production run of any Maxim gun. The gun itself weighed 23.8 kg/52.5 lb, an interesting comparison with the Vickers, and was generally used on a small carriage which with its shield weighed 45.2 kg/99.7 lb. The caliber was 7.62 mm/0.3 inch, and fed from 250-round fabric belts the cyclic rate of fire was between 520 and 600 rounds per minute, slightly more than the Vickers'. The PM10 used the original Maxim toggle lock, as shown by the depth of the weapon below the bore where the Vickers was shallow.

GERMANY

The German Maschinengewehr 08 was also a Maxim gun and, like the Russian weapon, preserved the original layout of the toggle lock to produce a notably deep weapon. Caliber was the standard German 7.92 mm/0.312 inch, and the feed system was a 250-round belt for a cyclic rate of 300 to 450 rounds per minute, a rate kept deliberately low as the Germans were proponents not of massive firepower as such, but rather of the greatest possible quantity of aimed (and thus genuinely effective) firepower. This eased the problem of ammunition supply and barrel changing with-

The Besa machine-gun was produced in the UK from a number of Czech designs, but found more favor as the secondary armament of armored fighting vehicles than as an infantry light machine-gun.

out any real penalty in terms of tactical capability. Generally known as the Spandau after the arsenal in which it was made, the MG08 weighed 62 kg/136.7 lb complete with tripod mount and spares. Like the Russians, the Germans provide this piece of "light artillery" with a sledge mount for tactical mobility. German machine-gunners were picked men, the army command appreciating by late 1914 that the machine-gun was king of the battlefield. These men were notable for their dedication and skill, as the dreadful British and French losses in battles such as Chemin-des-Dames, Loos, Neuve Chapelle and Champagne fully testify.

Though the Vickers, PM10 and MG08 were all variations of the same basic design, it is interesting to note that the Vickers had a muzzle velocity of 2440 ft/744 m per second with a 28.4-inch/721 mm barrel, while the comparable figures for the MG08 were respectively 863 m/2831 ft per second with a 720 mm/28.35-inch barrel and 900 m/2953 ft per second with a 719 mm/28.3-inch barrel.

AUSTRO-HUNGARY

The Austro-Hungarian heavy machine-gun was the Schwarzlose Maschinengewehr Model 1912 in 8 mm/0.315-inch caliber. This water-cooled delayed-blowback weapon weighed 19.9 kg/43.9 lb and was generally used on a 19.8 kg/43.7 lb tripod mounting. The weapon worked well enough, but the barrel was slightly too short at 526 mm/20.7 inches for the power of the round used, with the result that the Schwarzlose frequently revealed its location through bright muzzle flash. Feed was by 250-round fabric belts, the muzzle velocity was a poor 620 m/2034 ft per second, and cyclic rate of fire 400 rounds per minute.

USA

Notwithstanding its demands for French Hotchkiss machine-guns, the US Army did use a number of American machine-guns, including a small number of the Colt-Browning Model 1895 air-cooled gas-operated weapon that led eventually to the refined Marlin machine-gun. Available in larger numbers and generally far more satisfactory was the Machine-Gun M1917, another Browning design. This weapon was based on the short-recoil system, however, and paved the way for a series of great Browning machine-guns. Like the Model 1895 it had a distinctive US design feature, namely the pistol grip handle (at a time when most European weapons used one or two spade grip handles) at the rear of the weapon, with a trigger just in front of it. Browning had started work on the new gun as the Model 1895 entered production, but there was no official interest in the new weapon as money was short and the Model 1895 was considered adequate for the 'colonial' type of warfare envisaged by the US Army at the beginning of the 20th century. This all changed as the US entry into World War I approached, for it was woefully equipped, especially with automatic weapons, for the static warfare in the European theater. Despite the speed with which the type was rushed into production and service, the M1917 proved itself remarkably trouble-free, an ideal testimony for that great designer John Browning. The type was chambered for the standard 0.3-inch/7.62 mm US round, developing a muzzle velocity of 2800 feet/853 meters per second with the M1917's 23.9-inch/607 mm barrel, and weighing 32.6 lb/14.8 kg without water, the M1917 was competitive with European machine-guns. The tripod weighed 53 lb/24 kg, and firing from a 250-round fabric belt

Above:
A British stretcher party passes a French defensive outpost line during the great German offensives of early 1918. The machine-gun is an 8 mm Hotchkiss Model 1914 on the Model 1916 tripod mounting.

Top right:
In the period after World War I it was realized that the Lewis gun did not need its special air-cooling jacket, and reserve stocks of the type were modified accordingly to become useful light AA and second-line weapons during World War II.

Right:
In a typical propaganda photograph of the period, 2nd Lieutenant V.A. Browning fires an M1917 machine-gun, the type designed by his father, the great John Browning.

the M1917 had a cyclic rate of 450 to 600 rounds per minute.

Mounted semi-statically on their tripod mounts in carefully-sited fire positions, these and other heavy machine-guns proved themselves decisive in defensive warfare, and a useful adjunct to offensive warfare. However, attacking troops were all too aware that they were often pinned down by enemy machine-guns in spots where they lacked their own machine-guns to suppress the opposition. The policy of the period was to move up the heavy machine-guns only after an objective had been secured, as a means of defeating the inevitable counter-attack. Of course this begged the question of how the attacking infantry were to take the position in the first place: theory had it that the supporting heavy machine-guns could contribute as they had long range for accurate fire, but this ignored the practical difficulties of telling the machine-guns where to fire in a radioless tactical situation that had begun to diverge from the preordained scheme.

Light machine-guns

UK

By the end of 1915 the problem had become acute as trench warfare was firmly embedded in the soil of northern Europe. It was the British who took the lead, though the circumstances were economic rather than military, at least in the short term. The virtual destruction of the British regular army in 1914 and early 1915 had led to the raising of the vast "Kitchener armies", and the logistical problem thus raised for the British was how to equip them. So far as automatic weapons were concerned, the British opted for the Lewis gun, a light machine-gun which took only one-sixth of the time it took to produce a Vickers gun. Initial plans called for each battalion to have four Lewis guns as the Vickers weapons were withdrawn to equip the new Machine-Gun Corps, but as the real benefits of the Lewis became apparent the establishment was doubled to eight in mid-1916 and a few weeks later raised to 12, and by the

The 0.303-inch Lewis light machine-gun with overhead rotary magazine.

year's end one Lewis was shared by every four platoons, and by November 1918 two platoons.

The Lewis gun was in fact designed by Samuel Maclean, but the rights were then sold to another American, Colonel Isaac Lewis, who promoted the type in the USA and Europe from 1910. Rights were taken up in the USA by the Automatic Arms Company of Buffalo, New York (though US official interest was minimal) and in Europe by Fabrique Nationale of Belgium. Belgian manufacture began in 1913, but in 1914 the Birmingham Smalls Arms (BSA) company took over, putting the company in an ideal position when the expansion of the British army began later in that year. In itself the Lewis was an unremarkable gas-operated weapon with a complex mechanism that required a great deal of careful maintenance. But it was the weapon's comparatively low weight and feed system which made it so important. The gun used an ingenious forced-draught air-cooling system, and thus weighed a mere 27 lb/12.25 kg, while the ammunition feed was a 47- or 97-round rotary drum above the receiver. The magazine proved vulnerable in service, and even slight damage could cause jams. The Lewis's cyclic rate was 450 to 500 rounds per minute, and its muzzle velocity 2440 feet/744 meters per second with a 26-inch/661 mm barrel.

Despite these problems, the gun was enthusiastically received by the troops: here at last was a weapon light enough for attacking soldiers to take with them, providing them with their own fire support where and when they really needed it. It is fair to say that the bipod-mounted Lewis gun completely reordered the nature of tactics on the Western Front, and the ultimate accolade was that the Germans rapidly pressed all captured examples into service, while the Americans belatedly joined the production effort with a version chambered in the US 0.3-inch caliber. With the adoption of the Lewis and the realization of its capabilities by the Germans (who took to picking off the Lewis gunners before turning their attentions to the rest of the attacking force), the British evolved rushing tactics in

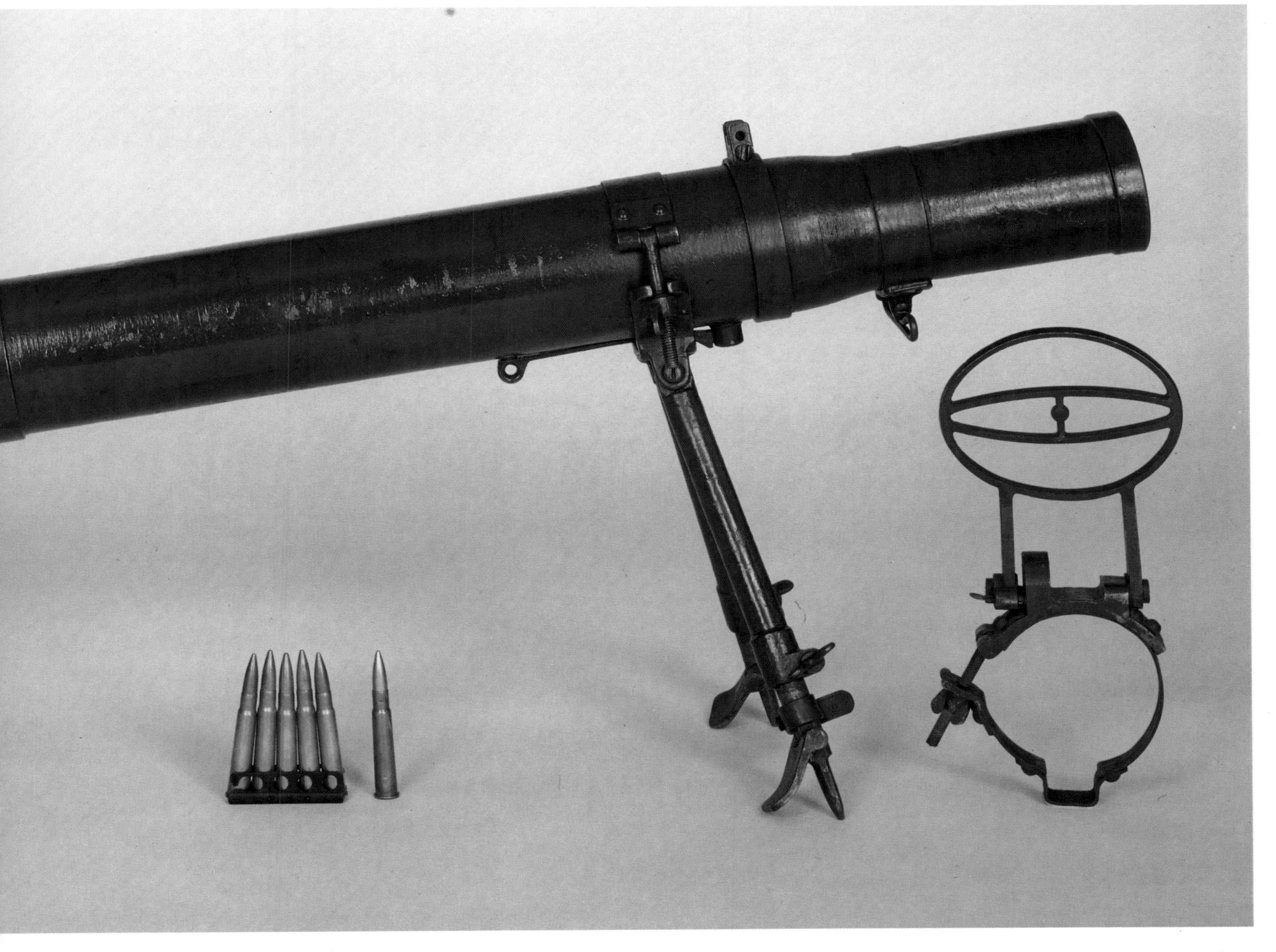

Left:
The Breda Model 1930 became Italy's standard light machine-gun in World War II, and was a thoroughly indifferent weapon requiring the use of a pump-action lubrication device to oil the rounds and so ease extraction. This added weight and complexity to the weapon, and the oil also picked up dirt and debris to jam the weapon.

Top right:
The Vickers machine-gun needed few modifications, though these weapons in service with the 5/5th Mahrattas in World War II have recoil boosters at the muzzle to improve reliability.

Bottom right:
Many expedients were tried to increase the mobility of water-cooled machine-guns, one being installation in motorcycle sidecars, as in this shot of a motor machine-gun company departing their camp on the Albert-Amiens road during late 1916.

place of the previous linear assaults: covered by a Lewis, a section or platoon rushed forward a short distance and then took cover so that it could provide support for the original support unit. The Germans quickly adopted the same tactic and after suffering its effect at Verdun in 1916, so too did the French. Mutual-support advance by smaller parties thus came to succeed mass attack as the primary infantry tactic.

GERMANY

In their usual far-sighted manner, however, the Germans had gone one step farther than the Allies: rather than use support tactics to capture the original objectives, the new German *Sturmtruppen* (storm troops) disregarded the original type of objective, instead using the rushing mutual-support tactic to sweep on past the trench system to the artillery lines in the rear and so destroy the basis of Allied defense. And the weapon the Germans evolved for this role was the Maschinengewehr 08/15 (MG08/15), a much lighter and modified version of the MG08, with consequent advantages in maintenance and spares holdings. Trials before the MG08/15's selection evaluated the Madsen, Dreyse and Bergmann light machine-guns, but the capabilities and logistic advantages of the MG08/15 swayed the final selection. Interestingly (and quite rightly) the Germans did not call the MG08/15 a light machine-gun, for it still weighed 18 kg/39.7 lb as it retained the MG08's water-cooling system, albeit with a smaller jacket. Major modifications were a bipod mount in place of the heavy tripod, lighter walls in the receiver, and a butt/pistol grip so that the weapon could be aimed from the shoulder. The standard 250-round belt was too long for many tactical applications, and so the MG08/15 was provided with 50- and 100-round belts. An alternative was the use of a side-mounted belt drum to keep the fabric and attached rounds out of the mud.

Perhaps of greater significance for the future, the Germans in 1918 introduced a further lightened model, the MG08/18, in which the water-cooling system was replaced by a light fretted air-cooled jacket. Only a few such weapons were used operationally, but troops were highly impressed with the extra weight-saving of this model, and the lesson was remembered for later exploration.

FRANCE

In the light machine-gun field the French used initially the Hotchkiss Model 1909, a lightened derivative of the Model 1900 medium machine-gun in the same 8 mm caliber which was heartily disliked for its disastrous feed mechanism, and the equally poor Chauchat, introduced in 1915. Disastrously designed by a commission, this was a thoroughly unreliable weapon adopted for production only because of vested interests. The weapon used the long-recoil operating system, in which the barrel and bolt moved together to the rearmost extent of the recoil, the bolt then being checked as the barrel was driven forward, following after a slight interval to strip and chamber the next round. Such a system involves the movement of much mass in the weapon (making aim difficult) and requires both careful manufacture and top-grade materials, both of which were skimped, and the result was a very poor weapon whose only attributes were a weight of only 9.2 kg/20.3 lb and a neat 20-round box magazine curved round into a semi-circle to save depth below the weapon.

USA

The US solution was an odd weapon which the Americans called an automatic rifle and all others a light machine-gun. Yet another of the great Browning designs, this was the Browning Automatic Rifle, almost inevitably called the BAR. Capable of single-shot or automatic fire, the gas-operated BAR was light enough at 16 lb/7.3 kg to be carried and operated by a single man. There was (at least in the World War I version) no bipod or even monopod, so the weapon had to be used as a rifle: it can perhaps be regarded as the progenitor of the assault rifle, for its cyclic rate was 550 rounds per minute and it was fed with 0.3-inch ammunition from a 20-round box magazine. The BAR was very highly regarded by US troops in both world wars, and it was certainly a strong, beautifully designed and excellently made weapon. The type reached its development peak in the later 1930s, when the BAR M1918A1 appeared with a neat folding bipod and other detail improvements.

In the event, the static domination of World War I's battlefields by the machine-gun and its barbed-wire outer defenses was broken by the tank. The European powers and the USA realized that the dreadful casualty rates of the Great War could be avoided by emphasis only on mobile warfare in which the tank and its derivatives (supported by aircraft) would prevent the resurgence of the machine-gun.

Medium machine-guns

However, the machine-gun still had a vital role to play and great efforts were made to refine the type in three main areas: lower weight, higher rate of fire and lower production cost. The result was a third type of machine-gun in addition to the magazine-fed bipod-mounted one-man light machine-gun for assault, and the belt-fed tripod-

Right:
The British SMLE series of bolt-action magazine rifles reached its apogee with the No.4 Mk I, a reliable and sturdy weapon well suited to the mass-production techniques of the period and also capable of rapid fire when required.

Below:
Undoubtedly the greatest machine-gun developed in the inter-war period was the German MG34, which combined low weight with excellent firepower to become a weapon capable of true general-purpose use. This example is fitted on the Lafette 34 tripod mount for the sustained-fire role in North Africa during 1942.

mounted squad heavy machine-gun for support or defense requiring sustained fire: the general-purpose or medium machine-gun. This combined attributes of both World War I types in that it was light enough to be used as a single-man assault weapon yet sturdy enough to be mounted on a tripod and used for sustained fire if necessary.

Light machine-guns generally followed the pattern set by World War I types such as the Lewis gun and MG08/18: they were usually magazine-fed (20 to 30 rounds), air-cooled, bipod-mounted, weighing slightly over 20 lb/9.1 kg and about 48 inches/1220 mm long. Typical of such weapons were the Czechoslovak vz26 and vz30 (both in 7.92 mm caliber), the disastrous Italian 6.5 mm Breda modello 30, the mediocre Japanese Type 11 and Type 96 (both in 6.5 mm caliber), the moderate French Fusil Mitrailleur 1924/29 and Mitrailleuse 1931 (both in 7.5 mm/0.295-inch caliber), the magnificent British 0.303-inch Bren gun, and the highly capable Soviet 7.62 mm Degtyarev DP. Of these the best was undoubtedly the Bren gun, developed from the Czechoslovak vz33 produced at Brno by the Royal Small Arms Factory in Enfield, hence the designation. The Bren gun was gas-operated, extremely reliable and very accurate in either single-shot or automatic fire, and was capable of a 500-rounds-per-minute cyclic rate of fire. Feed was from a 20-round curved overhead box, and firing the standard British 0.303-inch round the 25-inch/635 mm barrel produced a muzzle velocity of 2440 feet/744 meters per second, sufficient to give the weapon excellent stopping power at medium and even long ranges. Yet at just over 22 lb/10 kg and 45.5 inches/1156 mm long the Bren gun was an ideal light assault weapon fired either from the hip or on its bipod mount of the typical folding type.

Heavy machine-guns followed the pattern of World War I, and indeed the same weapons were frequently used. However, two outstanding newcomers were the American M2 and the Soviet DShK, both 12.7 mm/0.5-inch air-cooled weapons with great range and general firepower (the M2's stopping power against a human target may be gauged from the fact that it can also be used against light armored vehicles). The M2 was a Browning design, and first appeared in 1921. The truly outstanding qualities of this weapon are attested by its still being in production some 65 years later. The core of the M2 is its special round, which was developed on the basis of the special 13 mm/0.51-inch anti-tank round used in the German World War I T-Gewehr anti-tank rifle. The original M1921 version was essentially a scaled-up and air-cooled M1917 machine-gun, and this holds true of the later M2 and M2HB heavy-barrel versions. The weapon uses a 110-round metal-link belt of the type that became increasingly common in World War II as it was relatively immune to water and mud, made ammunition positioning easier, and was also less prone to twisting. With a 45-inch/1143 mm barrel it has a muzzle velocity of 2900 feet/384 meters per second and a cyclic rate of 450 to 575 rounds per minute. As might be expected of so powerful a weapon, the gun is heavy, weighing 84 lb/38.1 kg, though the M3 tripod mount is surprisingly light at 44 lb/20 kg.

The DShK entered service in 1938 and, while not as impressive as the Browning in appearance, has roughly comparable performance. The gun itself is slightly lighter at 33.3 kg/73.5 lb, but was most often found on the massive wheeled carriage of the PM10.

Far left:
The Canadian Ross rifle was planned as a major service type, and was an excellent sniper's rifle. However, it was also too light and damage-prone for introduction as a standard infantry rifle.

Left:
The vast expenditure of machine-gun ammunition in World War I led to the development of specialist belt-filling machines that loaded the rounds accurately into the belt pockets far more effectively than men could manage.

Below:
The advent of Allied tanks prompted the Germans into swift development of a counter, one of the first being the massive T-Gewehr single-shot breech-loading rifle, whose round provided the design basis for the cartridge developed for the Browning M2 0.5-inch machine-gun.

Developed in the inter-war period, the Bren gun was the definitive light machine-gun of its era for its reliability, accuracy and high rate of fire. The overhead magazine meant that the weapon and gunner could keep as low as possible, giving them considerable tactical advantages.

USA

The most important of the new breed of general-purpose medium machine-guns were the American M1919 and the German MG34. The M1919 was essentially a reworking of the M1917 with an air-cooled barrel, which helped to reduce weight to 31 lb/14.1 kg for the definitive M1919A4 variant that was produced in vast numbers during World War II. The original intention for the weapon was as a machine-gun in the tanks which the Americans planned to produce in huge quantities during 1919, but the M1919's overall superiority to the M1917 led to its widespread adoption as a medium machine-gun. The standard 0.3-inch US round was fired from a 250-round belt, and with a 41-inch/1041 mm barrel the muzzle velocity was 2800 feet/854 meters per second. The M1919 had a useful rate of fire of 400 to 500 rounds per minute, but is most remembered as the standard US medium machine-gun of World War II and the Korean War, in which the type served faithfully and with utter reliability.

The role of machine-guns in World War I was opposition to or support of the infantry who had the grim task of emerging from their trenches, crossing no-man's-land and any wire entanglements that remained uncut by the preceding artillery bombardment, and then fighting their way into the enemy's trench system. With the trench system captured, the victors' machine-guns could then be brought up and re-sited as defensive weapons against counter-attack. But the infantryman's personal weapon was the rifle, which had reached a peak in the years before World War I but then proved generally unsuitable for the trench warfare of that conflict. The peak was the manually-operated breech-loading magazine rifle, which had been designed towards the end of the 19th century as the accurate weapon of regular infantry, who theoretically would have the musketry skills to pick off open-order attackers at long range from prone or kneeling positions. This emphasis on musketry had resulted in precision weapons with long barrels and high muzzle velocities, which were almost a hindrance in World War I.

Far left, top:
The Japanese Type 92 machine-gun seen here in action in China during 1932 was a standard Hotchkiss type, easily recognizable by the cooling rings on the barrel. The type was widely used, but its weight did not lend itself well to the fast-moving operations favored by the Japanese.

Far left, bottom:
Classic stuff from 1945: a BAR opens up in support of a rifleman with a Garand M1.

Right:
Typical of the propaganda work of World War I, this 1916 photograph shows the knurled grips of the Luger pistol's toggle action, the implication being that the Canadian infantryman has taken the weapon from his prisoner—although the "German" is also holding a self-loading pistol, in this instance a Mauser.

Below:
The US equivalent to the British Lee-Enfield and German Gewehr 98 was the Springfield M1903, seen here in the hands of the first American contingent to arrive in the UK during 1917.

1915

After the opening campaigns of the war, in which the 19th-century musketry theory seemed to have been vindicated in battles such as Mons (where the high rate, accuracy and long range of the defending infantrymen persuaded the attacking Germans that the British must have equipped their infantry with additional machine-guns), the stalemate of trench warfare dictated new priorities in rifle design. Ruggedness to keep on operating under the worst conditions (and indeed to be used as a club at times), a moderately high rate of fire in the hands of semi-skilled soldiers, accuracy to 400 yards/365 m rather than the pre-war norms of 800 yards/730 meters or more, and shortness so that the weapon could be used easily in the confines of the trenches became priorities.

UK

The weapon that most nearly met these requirements was the British Rifle No.1 Mk III, the latest version of the Short Magazine Lee-Enfield Rifle introduced in 1907 as successor to the original Lee-Enfield rifle. During the Boer War the British discovered that the standard rifle was too long for effective cavalry use. The SMLE was thus a hybrid weapon weighing 8.7 lb/3.9 kg and 44.6 inches/1.13 m long. It was chambered for the powerful 0.303-inch round which, with a barrel length of 25.2 inch/640 mm, developed a muzzle velocity of 2080 feet/634 meters per second. Unlike the Mauser rifles, which used front-lug locking for the turn-bolt action, the Rifle No.1 had rear-lug locking which was in theory weaker but was trouble-free and gave an exceptionally smooth action that permitted very high rates of aimed fire with the 10-round box magazine, which was fed from above with ammunition in five-round charger clips. Various modifications led to the more easily produced Rifle No.1 Mk III*, which did away with some of the No.1 Mk III's long-range and precision features without lessening its utility in the trenches. The No.1 Mk III* was "the" British rifle of World War I, and was so successful that it remained in service throughout World War II together with the No.4 Mk I improved version.

USA

The No.1 was far better than any other Allied rifle, the best of which was the standard American rifle, the Springfield Model 1903 in 0.3-inch caliber. As its name indicates, this weapon was introduced in 1903 when the US Army realized that its 1892 Krag-Jorgensen rifles were becoming outmoded. In a comparison of contemporary designs the Mauser action was deemed the best, and license-production of a Mauser-type weapon was started at the Springfield Arsenal. The resulting Model 1903 was, like the SMLE and No.1, a compromise weapon with a 24-inch/610 mm barrel between that of early rifles (typically 30 inches/762mm) and current carbines (less than 22 inches/559 mm). The Model 1903 was nevertheless a fine rifle notable for its accuracy (to target-shooting standards and originating in part from the weapon's nice balance) and easily-used bolt action. Its overall length was 43.2 inches/1097 mm, weight 8.7 lb/3.9 kg, muzzle velocity 2800 feet/853 meters per second, and feed a five-round box.

Trench warfare demanded all their reserves of strength and determination from defenders and attackers alike. Here Canadians hold a German assault near Ypres during World War I in the classic way: a Vickers machine-gun supported by rapid rifle fire and grenades checks the Germans as they stumble through the killing zone impeded by barbed wire, imperfectly cut by the preceding artillery barrage.

FRANCE

The French used the Lebel Model 1886 in 8 mm caliber, but this was at best obsolescent with its tubular magazine with slow loading and the perpetual threat of explosion and 1303 mm/31.4-inch overall length. Other French service weapons of this basic type were the Berthier and Gras carbines (short and light, but with only small magazines and excessive muzzle flash as the standard 8 mm round was used with the 450 mm/17.7-inch barrel) and the Berthier Model 1907 rifle with the Lebel's action and a three-round box magazine.

BELGIUM, USSR AND ITALY

The Belgians used the 7.65 mm/0.301-inch FN-Mauser Model 1895 with a five-round box magazine, and the Russians had the sturdy 7.62 mm Mosin-Nagant Model 1891 with a five-round magazine. Both were unexceptional but adequate weapons other than their considerable overall length, especially when fitted with their long bayonets. The same may be said of the Italians' standard weapon, the 6.5 mm/0.256-inch Fucile modello 91, otherwise known as the Mannlicher-Carcano and fitted with a six-round box magazine. This Italian rifle was decidedly underpowered, firing a small bullet with a muzzle velocity of only 630 meters/2067 feet per second.

GERMANY AND AUSTRO-HUNGARY

The Central Powers fielded two excellent rifles: the German Gewehr 98 and the Austro-Hungarian Mannlicher Model 1895. The Gewehr 98 was a Mauser design, and was possibly the most significant and influential bolt-action rifle ever made. Generally well made, even in the difficult later days of World War I, it was of unexcelled strength as a result of its basic design and the use of a three-lug front-locking system for the straight-pull bolt. This action was not as smooth as the rear-lug type used in the SMLE and its derivatives, but was adequate for standard infantry use even if it reduced the rate of fire, especially as the charger-fed integral magazine held only five rounds. These are quibbles about an otherwise classic weapon of enormous structural and dynamic strength. Given the conditions of World War I, the main failing of the Gewehr 98 was its overall length of 1250 mm/49.2 inches. Firing the powerful 7.92 mm/0.312-inch German round, it had a muzzle velocity of 640 meters/2100 feet per second with a 740 mm/29.1-inch barrel, and weighed 4.2 kg/9.3 lb. The Gewehr 98 was also the starting point for the Karabiner 98a, introduced in 1904 and made in vast numbers especially after trench warfare experience persuaded the Germans that a shorter weapon than the Gewehr 98 was needed in the front line: the weapon was 1100 mm/43.3 inches long with a

Below:
Arduous training under combat conditions is one of the few ways to evaluate major weapons effectively, and the Rifle M1 used by this US Ranger is clearly getting the treatment—confirming the excellence of the basic Garand self-loading action.

600 mm/23.6-inch barrel, but otherwise retained the characteristics of the basic Gewehr 98.

The Mannlicher Model 1895 (Repetier Gewehr 1895) was in many respects comparable to the Gewehr 98, and was also a straight-pull bolt-action rifle with integral five-round box magazine. The weapon weighed 3.8 kg/8.3 lb and was 1270 mm/50 inches long overall, the 765 mm/30.1-inch barrel firing standard 8 mm ammunition with a muzzle velocity of 620 meters/2,034 feet per second. Like the Gewehr 98, the Model 1895 also spawned a carbine version, the Repetier Stutzen Gewehr 1895, which overcame many of the problems of the full-length rifle. The Mauser design provided the starting point for comparable rifles and carbines all over the world, and the Mannlicher design was extensively used in southern and eastern Europe as well as lending features to a host of other weapons.

1918–39

The development of rifles in the period between the two world wars followed a similar pattern of lightening, shortening and automating. The result was a generation of handier weapons with higher rates of fire, and these were optimized for the shorter battlefield ranges that had become standard in World War I. In some cases this meant little alteration to earlier weapons: the British, for

Left:
In its initial form the Browning Automatic Rifle lacked a bipod; here the assault method that was favored by US infantry is shown, with the weight of the weapon supported by a sling round the shoulders.

Below:
The Model 1891/30 rifle used by the Soviets in World War II was still an effective long-range weapon, much favored when fitted with the PE (x4) telescopic sight.

example, stuck with the basic Lee-Enfield, the Germans with the Gewehr 98 and the Soviets with the Mosin-Nagant. However, while the British No.1 Mk III was a good weapon with a higher-than-average rate of fire for a bolt-action weapon, it was expensive to make and so was redesigned as the more easily mass-produced No.4 Mk I, which was also a better weapon as it had a heavier barrel and longer-base sights (the rear sight was shifted back to a place over the receiver), both improving accuracy at all ranges. Production began in 1939, and amounted to an enormous number of weapons.

The Germans were satisfied with the Gewehr 98 except for its length, and thus in 1935 introduced the Gewehr 98k with a shorter barrel (600 mm/23.6 inches compared with 740 mm/29.1 inches) to reduce overall length from 1250 mm to 1108 mm (49.2 to 43.6 inches) and weight from 4.2 to 3.9 kg (9.3 to 8.6 lb).

USSR

The Soviets followed much the same road with the Mosin-

The rifleman in classic pose: a German infantryman moves low and fast with his Karabiner 98k.

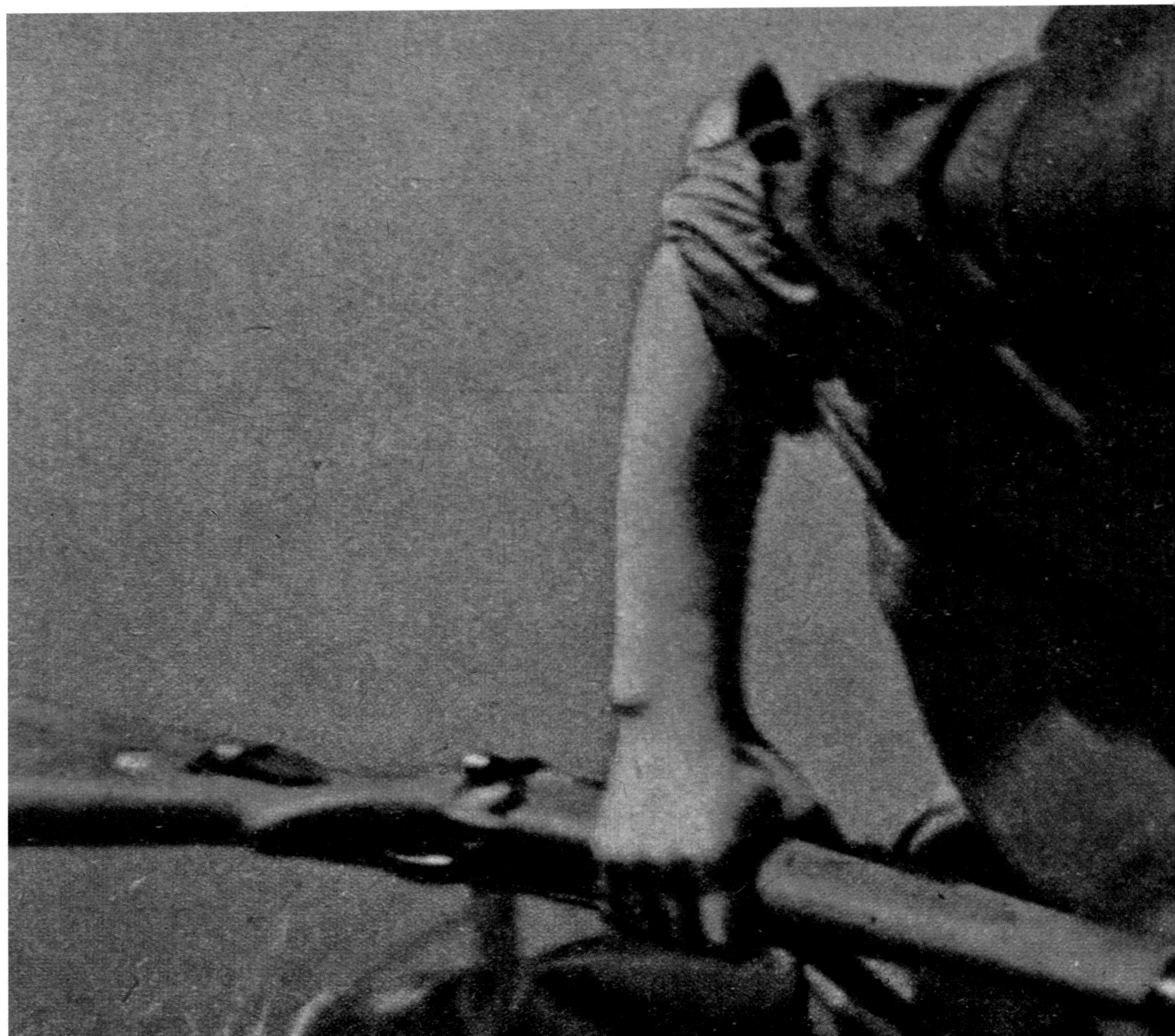

Nagant rifle, in 1930 introducing the Model 1891/30 with a 729 mm/28.7-inch barrel. The Soviets also introduced a shortened version of the Mosin-Nagant Model 1910 carbine, the Model 1938 with a barrel of 508 mm/20 inches. Both these weapons were rugged arms well suited to Soviet operational doctrines, but the extent of growing Soviet technical innovation is indicated by the adoption of a Tokarev-designed semi-automatic weapon. This was the gas-operated 7.62 mm SVT38 of 1938, which proved too flimsy for extensive Soviet service and was thus followed in 1940 by the more robust SVT40. Both weapons were nicely engineered, but suffered from the use of a powerful round in terms of heavy recoil and excessive muzzle flash.

USA

Self-loading also appealed strongly to the Americans, who in 1932 adopted the 0.3-inch Rifle M1, a semi-automatic weapon generally known as the Garand after its designer. This was an excellent rifle of its type, but complex in its gas operation and thus expensive. Nevertheless the Americans

Above:
Though unexceptional, the Russians' Mosin Nagant Model 1891 rifle was serviceable and capable of taking the hard knocks to which it was subjected. These Russian infantrymen are pictured at Salonika during 1916.

Left:
One of the great French disasters of World War I was the "Chauchat", otherwise the Fusil Mitrailleur Model 1915. Much had been expected of this light assault machine-gun, but it was designed by a second-rate committee and emerged as a distinctly poor weapon that was put into production to satisfy the demands of French arms manufacturers.

produced the weapon in vast numbers with few modifications, and it served them well right up to the 1960s. The weapon weighed 9.5 lb/4.3 kg and was 43.6 inches/ 1107 mm long overall. The barrel was 24 inches/609 mm long, producing a muzzle velocity of 2805 feet/855 meters per second. The clip-fed five-round box magazine held only eight rounds, and the empty clip was ejected very noisily.

Another US semi-automatic weapon was the odd Carbine M1, a short and usefully light weapon firing a unique intermediate-power standard 0.3-inch caliber round. The weapon was gas-operated and used a 15- or 30-round box magazine, but despite a low muzzle velocity of 1970 feet/ 600 meters per second and very limited effective range of 110 yards/100 meters, the M1 attracted great front-line favor from men who liked its low weight of 5.2 lb/2.36 kg, short 35.6-inch/904 mm length and good ammunition load. The weapon had been designed for second-line troops and officers, but was soon very widely used.

FRANCE AND JAPAN

The French used their elderly Lebel and Berthier rifles, and also the new and generally unsatisfactory 7.5 mm MAS36 rifle, and the Japanese had the 7.7 mm bolt-action rifle introduced in 1937 as a re-calibered 6.5 mm Type 38 rifle, which had been introduced in 1905.

Above:
Bulgaria's standard rifle of World War I was the Mannlicher Model 95, seen here resting on the parapet of a well-prepared Bulgarian defensive position.

1939–1986

There is a military truism that armies are at any time best equipped to fight their previous war, and this holds particularly true of World War II: a whole new series of weapons had been developed to meet the tactical situations of World War I. Some of the weapons developed did have considerable use, but other lessons had been misappreciated or mislearned, and other weapons were not suitable for a war whose campaigns were deliberately kept as fluid as possible. So far as the infantryman and his weapons were concerned, World War I had shown that combat ranges were decreasing, and the small arms of the 1920s and 1930s fully reflected this. World War II showed that these ranges were still decreasing, and that the individual infantryman needed to generate a higher volume of fire either to keep the heads of the enemy down, or to stand a realistic chance of hitting a fleeting target.

It is worth emphasizing here that the steady linear advances of troops common in the 19th century could be met by perhaps 10 aimed rounds from the muzzle-loader of each defending infantryman. The legacy of this was the mayhem created by the British Expeditionary Force against the Germans in the Battle of Mons in 1914, when high-volume aimed rifle fire took so heavy a toll that the Germans thought they were faced with a concentration of machine-guns. The decreasing combat ranges of World War I after the onset of trench warfare still allowed the defenders (with the support of machine-guns) to loose a lot of aimed fire as the attackers crossed no-man's-land, which led to the evolution and widespread adoption of light machine-guns and mutual-support tactics. However, this

Below:
The M16/M203 combined assault rifle/grenade-launcher in action during a US training exercise in the desert.

Left:
Belgian soldiers field-strip their FN FAL rifles during an exercise. Though heavy and expensive by modern standards, the FAL is still an attractive weapon for its durability and excellent stopping power.

type of small-unit tactic was developed so much in the 1930s and during the opening phases of World War II that the musketry chances of the defending infantry were greatly reduced without a high volume of fire. The attackers' side also demanded high volume of fire so that small infantry units could deal competently with the opposition after they had reached their final assault objectives.

Assault rifles *World War II*

This demand for more fire at shorter ranges left even the shorter rifles of World War I at a disadvantage: they were too heavy and because they fired a high-power round with a fairly massive bullet lacked the magazine capacity (even in the self-loading weapons) for short, sharp engagements. One exception was the US Carbine M1, which had been designed for second-line duties but was adopted in front-line units as ideal in the new type of warfare: the effective range might be only 110 yards/100 meters as a result of the low-powered round, but the weapon was handy and the magazine held a useful quantity of ammunition, reducing the need to reload. If reloading was necessary, it was possible to change the magazine very rapidly.

The German army, equipped with the 7.92 mm/0.312-inch Gewehr 98 and Karabiner 98k weapons at the beginning of the war, soon discovered this for themselves despite the tactical advantages of being the aggressors and using advanced operational techniques such as fast-moving armor in conjunction with dedicated close air support. It soon became clear that a self-loading infantry weapon was needed, and in 1941 the 7.92 mm Gewehr 41(W) was adopted. This was a Walther design based on the Danish Bang system using a muzzle trap for the gases needed to operate the piston system but it soon proved too complex for reliability in campaigns such as that on the Eastern Front. On the Eastern Front, however, the Germans encountered the Tokarev self-loading weapon, operated by gas tapped off through a hole in the barrel. This was manifestly an effective and reliable system, and the Germans copied it into the Gewehr 41 to produce the far more reliable Gewehr 43. This weapon was 1117 mm/44 inches long and weighed 4.4 kg/9.7 lb, but proved sufficiently accurate with a barrel only 549 mm/21.6 inches long to serve as a sniper's rifle. The magazine held 10 rounds, double the capacity of the Gewehr 98 series' magazine, and the muzzle velocity was 776 meters/2546 feet per second. The increasingly dire war situation later forced the Germans to economize, and the Gewehr 43 appeared with plastic rather than wooden furniture and with other production shortcuts.

Yet the Gewehr 43 and even shorter Karabiner 43 were not the answer to the Germans' need for higher rates of fire. Better results were achieved with the limited-production Fallschirmjäger 42, designed by Rheinmetall for the Luftwaffe's paratroop arm. This used the standard

An evolutionary development of the M1 Garand self-loading rifle, the M14 is chambered for the 7.62 mm NATO round, and is a selective-fire weapon with a 20-round detachable box magazine. It arrived shortly before the US switched to the 5.56 mm caliber, but still serves with second-line, reserve and Army National Guard units.

Left:
The Colt Commando 5.56 mm assault carbine/sub-machinegun in action with a member of the 4th SEAL Team of the US Navy.

Below:
The German MG42 has been the most significant machine-gun since the Maxim gun appeared in 1885; a combination of low weight, exceptionally high rate of fire, considerable accuracy and range, and great reliability were all attributes of this great weapon, as well as low production cost and simple maintenance. Most modern machine-guns owe something to this seminal weapon.

7.92 mm round and had no completely new technical features, thus achieving good reliability right from the start: but where it was different was the combination of automatic characteristics, complete with a folding bipod into a highly compact weapon with a straight-through design from muzzle to butt. At the cost of difficult and thus expensive manufacture the FG42 achieved a selective-fire capability (single-shot, or fully automatic at a cyclic rate of 750 to 800 rounds per minute) with a detachable box magazine holding 20 rounds, all for a weight of 4.5 kg/10 lb and a length of only 940 mm/37 inches. It was a remarkable and prescient weapon that showed what was possible for the future once mass-production capability had been designed into such a weapon.

But given its cost and lengthy production time the FG42 could not be adopted for the army, which instead opted for a weapon designed for the new 7.92 mm short round developed by Schmeisser as an intermediate-power round best suited to the new type of assault rifle being advocated to replace the standard rifle. Hitler disapproved of the assault rifle concept, and the army had to prevent him from finding out what was afoot. The result was the Haenel-designed Maschinenpistole 43, the true precursor of today's assault rifles and a simple gas-operated weapon that was very reliable and cheap, as machinings were replaced as far as possible by stampings and pressings. The result was a thoroughly utilitarian weapon of decidedly aggressive looks, ideally suited to the requirements of the German army as it was forced increasingly onto the defensive in the second half of the war. The weapon delivered either accurate single shots for static defense, or fully-automatic fire for assault or close-quarter fighting. One of the primary advantages of the lower-powered round apart from its lower weight was the much reduced recoil, which made it feasible to fire the weapon on full automatic without the tendency of higher-powered weapons to muzzle-climb, so making accurate fire all but impossible for the first two or three rounds, after which the sky or the tree tops were the main recipients of the burst. Once the weapon was in production Hitler changed his mind about the type, and ordered it to be restyled the Sturmgewehr 44. The StG44 weighed 5.2 kg/11.5 lb, and the detachable curved box magazine held 30 rounds, sufficient for an effective burst at the cyclic rate of 500 rounds per minute. Length was a very handy 940 mm/37 inches, and as might be expected with an intermediate-power round the muzzle velocity was comparatively low at a modest 650 meters/2132 feet per second.

The tactical effect of the MP43/StG44 was enormous, for it removed from infantry the absolute need for machine-gun support. Such support was still useful (and under certain circumstances vital), but for close-quarter fighting the infantryman could from now on generate his own automatic fire. German tactics were immediately adapted to the new weapon, and so great was the demand that an enormous production programme was initiated, though supply could never match demand.

Moving cautiously through a Belgian village in 1944, this American assault team features a point man with an M1 sub-machinegun, and two follow-up men with M1 carbines. Both weapon types were well suited to urban combat.

Above:
US infantry train with the M60 machine-gun in its assault role, fired from a kneeling position with the right forearm resting on the right knee for extra stability.

Left:
The M14 selective-fire 7.62 mm rifle was standard issue to all US infantry formations in the 1950s and early 1960s, replacing the M1 as the basic full-power rifle; it was then supplanted in most units by the 5.56 mm M16 series assault rifle.

Post-1945

The assault rifle proper was born in World War II out of German operational analysis that combat was almost invariably at ranges of less than 400 m/440 yards, while the full-power rifle cartridges of the day were designed for ranges over 1000 meters/1095 yards. The result was a new Polte-designed short round in the Germans' standard 7.62 mm caliber, and as this was optimized for 400-meter combat its propellant load was sufficiently light to make possible fully-automatic weapons whose recoil could be handled easily in a one-man weapon. Another advantage of the smaller round was that the individual soldier could carry more ammunition. As noted above, the first full-production assault rifle series was the MP43/StG44 series, whose capabilities so impressed the victorious Allies that they started work on their own assault rifles, though there was at first considerable feeling that a need remained for the standard infantry weapon. The short-term result was a series of comparatively light and short definitive self-loading rifles, but still chambered for a full-power round carried in useful numbers in a large magazine. Some of these weapons were capable of automatic fire, but most operators decided that single-shot (or at most limited-burst) firing was adequate.

As with the new breed of general-purpose machine-gun, one of the first into the field was FN in Belgium, which introduced its Fusil Automatique Leger (FAL) during 1948: the weapon was initially chambered for the 7.92 mm x 33 short round, but early moves towards a standard NATO round produced a revision to the 7.62 mm x 51 round. The FAL soon found a ready market for Belgian and indigenously license-produced models, and has become one of the most widely used weapons ever developed. In some respects the FAL is an unusual hybrid, for though its gas operation (using gas tapped off the barrel through a regulator) and 20-round detachable box magazine were advanced at the time of the weapon's introduction, the high quality of the materials and overall finish were legacies of a more quality-conscious era. The weapon is 1143 mm/45 inches long and weighs 5 kg/11 lb loaded, and the normal muzzle velocity is 838 meters/2750 feet per second. In single-shot mode the FAL can get off 30 to 40 rounds per minute, while the cyclic rate of the automatic model is 650 to 700 rounds per minute.

Given the modern tendency towards smaller calibers, FN has developed a 5.56 mm version as the FNC, and it is likely that satisfied FAL customers may well turn to this new model as the 5.56 mm round reaches remoter corners of the world.

Other notable Western rifles of this genre are the

Heckler and Koch G3 and the Israel Military Industries Galil, both in 7.62 mm caliber. The G3 was developed from a Spanish CETME design and issued to the West German army from 1959 and later bought (from West German or local license production) for many other armies. The weapon is based on a roller-locking delayed-blowback action and was designed for mass production. It thus lacks the solid feel of the FAL, though it is in no way weaker, as it uses steel pressings and plastic items wherever possible. The definitive assault-rifle version is the G3A3, which is 1025 mm/40.35 inches long and weighs 5.025 kg/11.1 lb when loaded with a 20-round detachable box magazine. With a 450 mm/17.7-inch barrel the G3A3 has a muzzle velocity of some 800 meters/2625 feet per second, and on automatic fire has a cyclic rate of 500 to 600 rounds per minute.

The G3's being a generation later than the FAL is indicated by its more advanced design and construction and also by the diversity of the weapons in the same basic family. For example, there are carbine versions (some with barrels so short that the weapons are sub-machineguns in all but name), light machine-gun models (fitted with a bipod and heavier barrel, such as the HK21 belt-fed model that can also use the standard 20-round box magazine), an airborne G3A4 version (with a telescopic butt that retracts to each side of the receiver) and a sniper model. Some 48 countries use the G3 series, of which no fewer than 12 produce it under license. The type's one main drawback is use of the powerful 7.62 mm x 51 round, and the manufacturer has thus developed the HK 33 model chambered for the new 5.56 mm round. The HK33 can use 20- or 40-round box magazines, and is also available in the same range of alternative models as the basic G3.

The Galil uses the same operating system as the Soviet

Right:
With the designer watching attentively, an American parachute officer puts the Johnson Model 1941 light machine-gun through its paces on the range. This 0.3-inch weapon was recoil-operated and produced in small quantities for US special forces.

Below:
The selective-fire 7.62 mm vz58 assault rifle produced in Czechoslovakia looks like an AK-47, but is in fact quite different in its internal workings.

The Soviet equivalent to the air-cooled Browning M1919 and MG42 was the 7.62 mm SG43, an air-cooled weapon of great reliability and useful mechanical simplicity. The gun weighed 13.8 kg/30.4 lb and fired at 500 to 640 rounds per minute from 50-round metal-link belts. Where appropriate the carriage of the PM10 heavy machine-gun was used for tactical mobility.

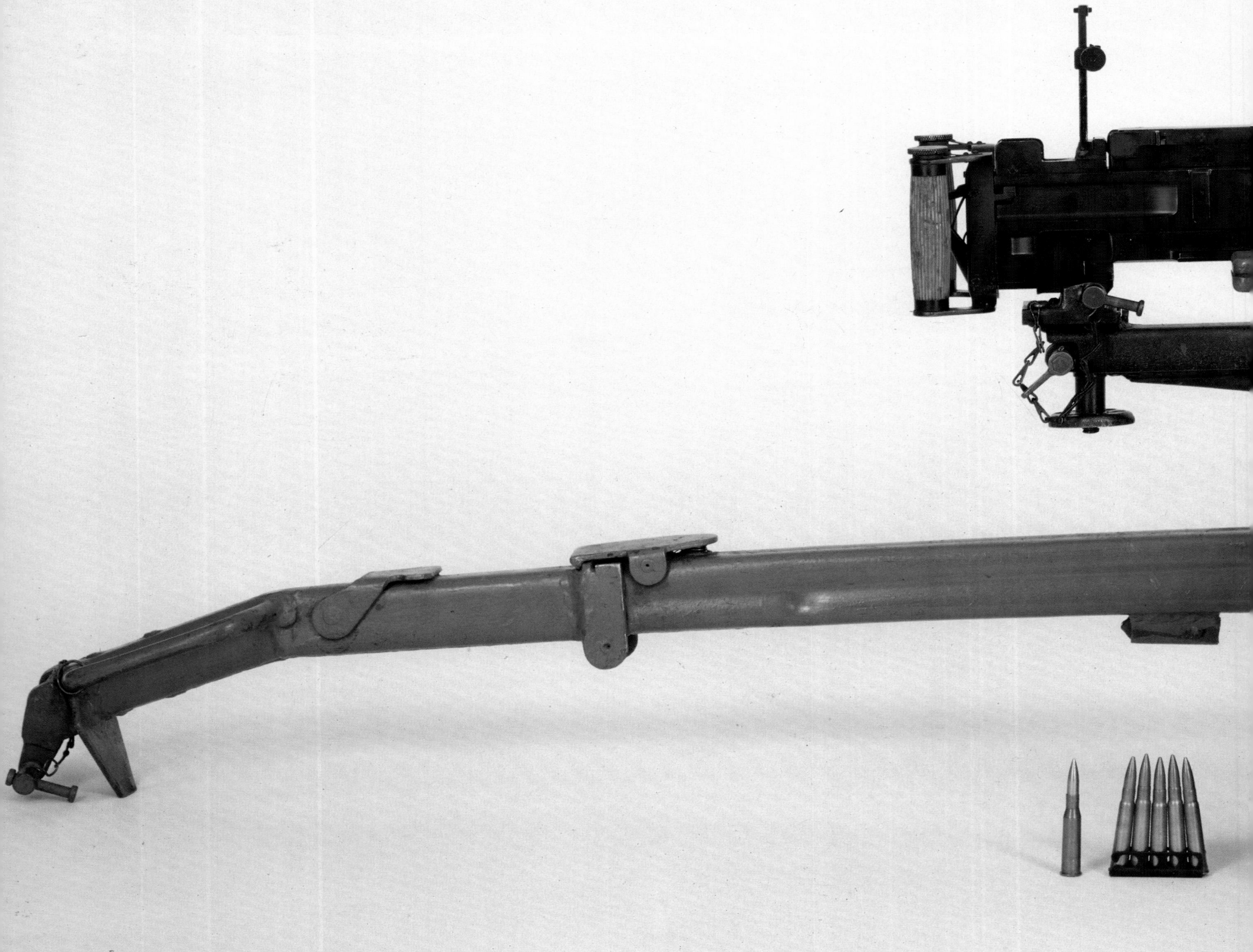

AK-47, from which it was derived via the Finnish Valmet assault rifle: gas tapped off the barrel works a piston moving the rotary-locking bolt. The weapon is widely used by the Israeli forces, is license-produced in South Africa as the R4, and formed the design basis of the Swedish FFV 890C. Simple yet highly effective, the Galil was originally produced in 7.62 mm caliber as the Galil ARM multi-purpose weapon with a bipod and carrying handle, the Galil AR assault rifle without the bipod and carrying handle, and the Galil SAR short assault rifle with a shorter barrel but otherwise similar to the Galil AR. All three models are fitted with folding stocks, and the practicality of Israeli weapon design is exemplified by the built-in bottle opener, fitted to prevent soldiers misusing vital features such as the magazine lips, and the wire-cutter on the Galil ARM's bipod. All models, as with other modern assault rifles, have a built-in device for launching rifle grenades, using a special 10-round magazine instead of the normal 20-, 35- or even 50-round magazines, though the latter two are used mainly with the Galil ARM for squad support. IMI has also followed the downward trend in assault rifle calibers, and the Galil is now produced in 5.56 mm form: in the ARM version the 5.56 mm model weighs 4.62 kg/10.2 lb with a 35-round magazine, the comparable weight for the 7.62 mm model being 4.67 kg/10.3 lb with a 20-round magazine. Both types have a cyclic rate of fire of 650 rounds per minute, but the 980 meters/3215 feet per second muzzle velocity of the 7.62 mm round drops to 850 meters/2790 feet per second for the 5.56 mm round.

On the other side of the fence the major weapon is the AK series designed by Mikhail Kalashnikov, undoubtedly one of the five most important weapon types ever produced. The series has been produced by the million, and can be found in every area of the globe where Soviet influence is felt, used by regular forces and also by "freedom-fighters" and their ilk. The AK series was clearly inspired by the StG44, and the original AK-47 model was designed for a 7.62 mm round equivalent to the Germans' 7.92 mm short round. The AK-47 and its 7.62 mm x 39 round were thus developed expressly in response to the threat of the StG44. The first AK-47 appeared in 1946, and has since become virtually a universal weapon. It is a simple assault rifle of great design clarity, made from high-grade materials but intended right from the start as a mass-production weapon. Overall length is a mere 869 mm/34.2 inches and weight 5.13 kg/11.3 lb with a loaded 30-round detachable curved box magazine, while the muzzle velocity is 710 meters/2330 feet per second with the 414 mm/16.3-inch barrel, and cyclic rate 600 rounds per minute.

So great was the success of the AK-47 (whose action was copied in many other designs, it should be noted) that by the late 1950s it was decided that the design should be updated to ease production. Thus there appeared the basically similar AKM with a stamped steel rather than machined receiver, and a simplified locking system. Apart from simpler manufacture, the primary difference is in ioaded weight, which drops to 4 kg/8.8 lb in the AKM. The

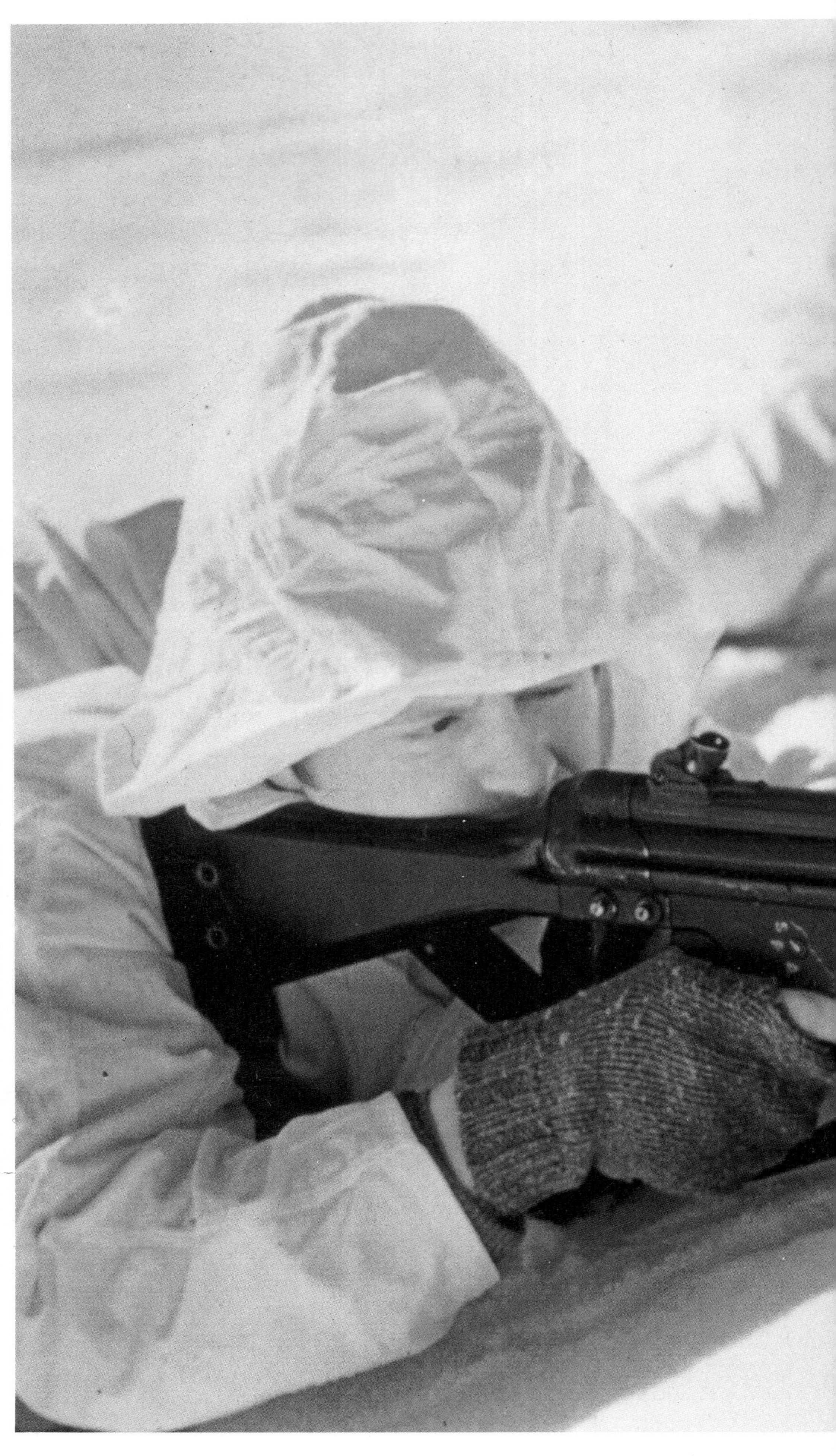

The Swedish AK4 is an all-but-identical licensed version of the magnificent G3 rifle made by FFV. The use of this rifle by the Swedes, who have to endure climatic extremes, is good testimony for the basic design's reliability.

A trio of East Asian weapons. From top to bottom: the Japanese Type 38 carbine, the Japanese Type 99 rifle and the Chinese Type 56 assault rifle, the last a copy of the Soviet AK-47.

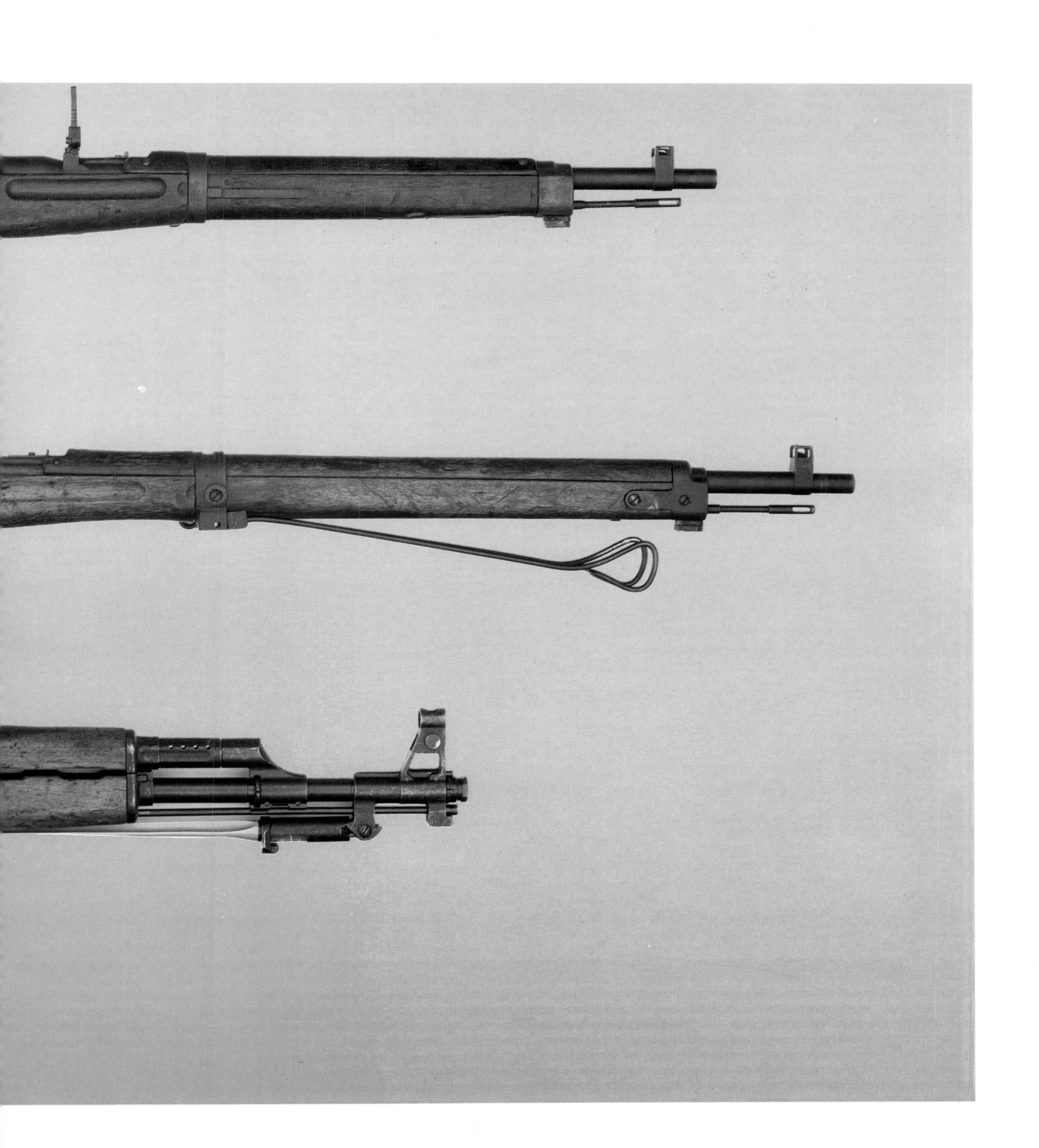

two types are still in worldwide complementary service, and will remain in such service for the foreseeable future.

Eventually the USSR decided to follow the West's lead into smaller calibers, and the result was the AK-74 firing the new 5.45 mm x 39 round. The AK-74 is the AKM redesigned as appropriate to take the new round, and first appeared in the early 1970s. The most noticeable external differences are a muzzle brake and a red plastic magazine that still holds 30 rounds but is shorter and less curved than the AK-47/AKM series'. The AK-74's loaded weight is 3.6 kg/7.9 lb and overall length 930 mm/36.6 inches, while the cyclic rate and muzzle velocity are 650 rounds per minute and 900 meters/2955 feet per second. As with the AK-47/AKM series, the AK-74 has been produced with solid wooden stocks and with folding metal stocks. Another point worth making about the AK-74 is its use of a bullet type prohibited under international law: this bullet has a hollow tip and a steel core with a center of gravity towards the rear of the bullet, resulting in tumbling as the bullet enters flesh. The effect is a particularly nasty wound. It should also be emphasized, however, that other rounds (such as the US M193) have the same tendency as a side effect of their shape.

The US M16 assault rifle in 5.56 mm caliber was adopted in 1961, and has its origins in a commercially-developed weapon, the Armalite AR-15 designed by Eugene Stoner. The AR-15 was essentially the 7.62 mm AR-10 revamped for the new 5.56 mm Fireball round (later standardized as the M193), and was bought by the US Air Force and the UK in 1961. The weapon is a gas-operated type with rotary locking, and is straightforward to make through the use of pressings and of nylonite for the furniture. The weapon became the M16 when the US Army adopted it, and since then has been produced by Colt in the USA and by overseas license-holders in the Philippines, Singapore and South Korea in vast numbers for the US forces and 12 other customers, mostly in the Far East, its 990 mm/39-inch length and 3.64 kg/8 lb loaded weight making it suitable for most East Asians' slighter build. The magazine is a 30-round box that has been adopted for several other weapon types, and other basic characteristics include a muzzle velocity of 1000 meters/3280 ft per second with the 20-inch/508 mm barrel, and a cyclic rate of 700 to 950 rounds per minute. The original M16 suffered from unreliable ammunition and bolt-closing difficulties in mud, the latter being solved by the addition of a plunger in the

Right:
The combined M16A1 rifle and M203 grenade-launcher is standard with the New Zealand Army.

Below:
The M14 rifle was standard issue at the beginning of the USA's involvement in the Vietnam War. The men of this 1966 patrol are equipped with the type, with the exception of the second man, who has an M60 machine-gun.

M16A1 variant 1966, and the latest model is the M16A2 with a heavier barrel for use with the new SS109 round. Light machine-guns with bipods, carbine and even a sub-machinegun have not achieved great currency. However, an addition that has become common is the M203 launcher for 40 mm grenades, fitted under the forestock and adding considerably to the versatility of the individual infantryman.

Machine-guns ***World War II***

Mass-production design was applied to the Maschinengewehr 34 general-purpose machine-gun, which was all that the German army could desire in the way of performance and reliability, but drastically expensive to produce. The possibility of long-term campaigns added urgency to the development of a cheaper replacement. The result was the Maschinengewehr 42, evolved by the Mauser designers

Above:
The Steyr AUG is one of the most advanced of the current generation of "bull-pup" weapons. The straight-line design reduces the tendency of the barrel to climb in automatic fire and also facilitates aiming.

via the MG39/41 on the basis of the MG34 with design input from Polish experimental weapons captured in 1939. In general, the MG42's operation was like the MG34's in that it was a strong and simple recoil-operated weapon with a muzzle trap to keep pressure high until the bullet was well clear of the barrel. Where the MG42 differed was in the bolt and locking system, the barrel and bolt recoiling in a straight line without turning while locked together. At the rear of the barrel was a screwed-on extension in whose sides cams were machined. As the bolt moved forward a locking stud on each side of the bolt hit the appropriate cam in the barrel extension, forcing roller locking lugs out of the the bolt into slots cut in the barrel extension as the bolt came flush with the rear of the chamber, so locking the bolt and barrel firmly together with a mechanically advantageous system. In the recoil phase the barrel and bolt

Above:
Excellently made and of first-class design, the Swiss SIG 540 series of assault rifles and sub-machineguns has not enjoyed significant sales, possibly because of their high cost.

A combined South Vietnamese and American team takes cover on the edge of a paddy field during operations in 1964. The weapons in evidence are M1 carbines and an M1919 medium machine-gun.

Above:
The Beretta AR70 is a neat assault rifle that has yet to secure the commercial success expected of such a cleverly-conceived design, made largely from non-strategic materials.

Left:
The new British assault rifle, the L85 Endeavour, is light and compact and offers formidable firepower at today's short battlefield ranges. The weapon's ability to accept a number of different sights greatly improves its tactical flexibility.

moved back together as long as there was a high gas pressure in the barrel. But as the rear of the recoil movement was approached, the studs on the bolt head were cammed out by the cams on the barrel extension, withdrawing the locking lugs and allowing the bolt and barrel to separate. A feed arm for the ammunition was also operated by the movement of the bolt, providing firm and completely accurate movement of the 50-round metal belt into a weapon with a prodigiously high rate of fire: at some 1500 rounds per minute, the MG42's cyclic rate was about double that of previous weapons, and made the weapon aurally most impressive, with a sound that has been likened to the tearing of linoleum. More to the point, the rate of fire made the MG42 a highly capable sustained-fire weapon with enormous impact in the shock phase of assault operations.

In themselves the high rate of fire and total reliability would have made the MG42 a noteworthy design. But additionally basic engineering development made the MG42 far easier, quicker and cheaper to produce than the MG34: sheet metal stampings were used wherever possible, notably for the receiver and the barrel housing, to ease production bottlenecks and speed construction. This use of metal stampings gave an angular utilitarian appearance, and made possible an ingenious barrel-change mechanism built into the barrel housing. Such changes were singularly important given the MG42's extremely high rate of fire. For assault the MG42 could be used from its folding bipod to deliver shock fire at close ranges, and a Lafette 42 tripod mount together with indirect-fire sights made the MG42 a capable or rather formidable sustained-fire weapon. Like the MG34, the MG42 was a highly capable light anti-aircraft weapon, and the Lafette 42 could be adapted simply to this latter role, again of increasing importance as the tide of war turned against the Germans and Allied tactical aircraft came to play an increasingly dominant part in the course of ground operations.

The MG42 was thus the most important machine-gun of World War II (and indeed of the period since the introduction of the Maxim gun), setting the pattern for the subsequent development of general-purpose machine-guns. The weapon fired the standard 7.92 mm round with a muzzle velocity of 755 meters/2475 feet per second through a barrel 533 mm/21 inches long. The MG42 was 1220 mm/48 inches long overall, and weighed a mere 11.5 kg/25.4 lb with its bipod. A true measure of the weapon's importance can be gauged from the number of derivatives it spawned after the war, and the extent to which its features were copied into other weapons. The type is still used by the West German army: the MG1 started to come from the Rheinmetall production line in 1959, and is virtually identical with the MG42 apart from using the standard NATO 7.62 mm/0.3 inch ammunition; the MG2 is ex-wartime stock rebuilt for the NATO round; and the MG3 is an updated version with a measure of light alloy (in less stressed parts) to reduce weight to 10.5 kg/23.15 lb. With its detachable butt the MG3 is 1225 mm/48.23 inches long, and the NATO 7.62 mm round produces a muzzle velocity of 820 meters/2690 feet per second with the 531 mm/20.9-inch barrel. The cyclic rate is variable between 700 and 1300 rounds per minute.

The weapon is also produced under license in Italy, Pakistan, Portugal, Spain and Turkey, and also without the benefit of a license in Yugoslavia, which adopted the type in 1945 and has since maintained the 7.92 mm caliber for its version, the SARAC M1953. French units took captured MG42s to Indo-China between 1945 and 1954, and the type has since been encountered in communist hands.

The MG42 was thus a seminal weapon in the development of modern general-purpose machine-guns, for Germany's opponents were made all too aware of the change wrought in battlefield tactics by this weapon to ignore the implications. The results were perhaps slow to materialize, but materialize they did and so helped to consolidate the wholesale change in tactics since World War II.

Post-1945

The result of World War II in the machine-gun field was a relatively slow spate of general-purpose machine-guns. Belgium was an early and highly successful entrant into this field, the Fabrique Nationale at Herstal producing the 7.62 mm Mitrailleuse d'Appui General (MAG) in the early 1950s. The weapon is of typical FN excellence in design and construction (some parts are sheet metal stamping riveted together, but most are machined high-quality steels), and is an unremarkable weapon except for the variable gas-regulator used to tap propellant gases from the barrel for the gas-operated action of the weapon. This regulator allows the machine-gunner to vary the rate of fire at will, and also makes possible continued operation in adverse conditions: for example, a fouled barrel can be overcome by opening the regulator to allow the use of more gas. The weapon also has an easily-changed 3 kg/6.6 lb barrel of the type vital for sustained fire, and can be fired from a light bipod or from a 10.5 kg/23.2 lb tripod mounting; in the latter role the butt is usually removed. The one failing of the weapon is its 50-round belt, which is perfectly acceptable for sustained fire, but is cumbersome and at times dangerous when the

The attributes of the M1 rifle were accuracy, reliability and considerable strength, all features that endeared the type to the American infantryman of World War II.

machine-gunner has to move rapidly in the support role (for which the weapon is also heavy by modern standards). The MAG is typical of its generation, with a weight of 10.1 kg/22.3 lb, an overall length of 1260 mm/49.6 inches, a cyclic rate of between 600 and 1,000 rounds per minute, and a muzzle velocity of 840 meters/2756 feet per second with a 545 mm/21.5-inch barrel. The MAG has been widely exported and looks set to remain in service far into the foreseeable future. Countries that make the MAG under license include Argentina, Israel, Singapore, South Africa and the UK. The British version is the L7 which incorporates a number of modifications to suit the weapon to specific British requirements.

The MAG is undoubtedly the most widely used post-war general-purpose machine-gun of Western origins, its only real competitor being the US M60, also a 7.62 mm-caliber gas-operated weapon. The M60 serves in large numbers with the US Army, and with a number of the USA's allies in eastern Asia, the most significant being Australia, South Korea and Taiwan; the last also makes the weapon under license. The M60 story began in World War II with the T44 development design, which capitalized on the many superior features of German machine-guns, including the piston and bolt of the Fallschirmjäger 42 and the ammunition feed system of the Maschinengewehr 42, allied to American-developed features in a design that made wide use of sheet metal stampings and plastics for ease of production and maintenance. After an extraordinarily protracted development and evaluation programme, the T44 became the M60 and was issued to US Army units in the late 1950s. Despite its lengthy gestation the new weapon was unreliable and unpopular, the major failing being the decidedly awkward barrel-change system, which required about half the weapon to be taken apart before the new barrel, weighing 8.245 lb/3.74 kg, could be fitted, clearly a situation fraught with tactical danger. Extensive redevelopment brought the weapon up to a fully acceptable standard even if it remained generally unpopular for its lack of balance. This is so pronounced that M60 machine-gunners generally carry their weapon with a sling rather than by the carrying handle, which is not on the weapon's center of gravity. The sling also means that the gunner can fire the weapon on the move, in much the same way as was pioneered by the BAR gunners of World War I before the type was provided with a bipod in its BAR M1918A1 form. For sustained fire the M60 is generally used on a tripod mounting.

Salient features of the M60 which, like the MAG, has been developed for use on armored fighting vehicles and helicopters (in remotely- or directly-controlled installations) include an overall length of 43.5 inches/1105 mm, a barrel length of 22 inches/559 mm, a weight of 23.17 lb/10.5 kg, a cyclic rate of 550 rounds per minute, a muzzle velocity of 2805 feet/855 meters per second, and 50-round metal-link belt ammunition feed.

The US Army plans to replace its M60s with the new M249 version of the Belgian Minimi, though the latter's caliber of 5.56 mm means that the M60 will possibly be retained as a medium machine-gun. A revised version currently being evaluated by several potential operators is the Maremount Lightweight Machine-Gun. This is a much improved weapon fully indicative of what the M60 could have been some 30 years ago with a little foresight: the weapon has been lightened and simplified (especially in the gas-operated part of the system), the bipod has been moved back to under the receiver, and a foregrip has been added. The lighter weapon is also considerably handier than the clumsy M60.

A growing tendency in recent years is the design and

Left:
The ultimate in current squad support weapons, the 5.56 mm FN Minimi can be fed from a metal-link belt, a magazine, or a box containing metal-link belt.

Below:
A US infantryman demonstrates the M14E1 7.62 mm rifle, a version of the standard M14. It has a wire-framed folding stock for a reduction in weight and for shorter overall length when required.

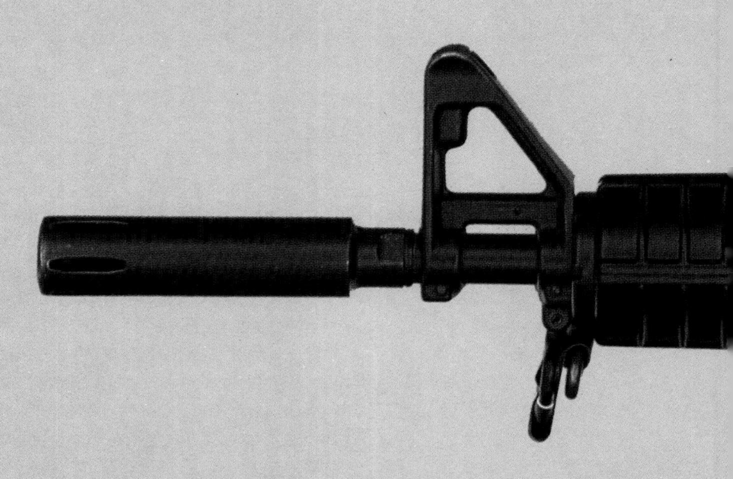

Three American weapons, from top to bottom: the M1A1 lightened carbine, the Armalite AR–15 (M16) and the Colt Commando carbine version of the M16 series.

COLT
AR-15
COLT
AR-15

manufacture of indigenous weapons outside the major power blocs, the most important region in this respect being South America. As such weapons can be imported without the political and/or economic strings attached to US or Soviet arms supplies, it is likely that weapons such as the Brazilian Uirapuru Mekanika will find a comparatively ready export market. The Uirapuru Mekanika is a fairly orthodox gas-operated general-purpose machine-gun of ungainly appearance, which can achieve a cyclic rate of 700 rounds per minute firing standard NATO 7.62 mm ammunition. With butt and bipod the weapon is fairly heavy at 13 kg/28.7 lb, but a moderate price should secure interesting export sales. The type can be used for support fire with its bipod, and in the medium role on a tripod and fitted with a heavier barrel. Ammunition feed is from a 50-round belt. If the Uirapuru Mekanika secures the anticipated export orders, there seems every likelihood that other South American small arms will also appear for local use and overseas sales.

On the other side of the Iron Curtain the major arms suppliers are Czechoslovakia and the USSR. The former's primary contender in the light machine-gun market is the vz59, a 7.62 mm weapon that appeared in the later 1950s as successor to the overly complex vz52, itself a belt-fed derivative of the pre-war magazine-fed weapons of the vz26 series. The type is very similar in its workings to the original models, but can be used for squad support with a bipod and light barrel, or as a medium machine-gun with a tripod and heavy barrel, in both cases with 50- or 250-round belt feed. The light barrel is 593 mm/23.35 inches long and produces a muzzle velocity of 810 m/2657 ft per second, while the heavy barrel is 693 mm/27.3 inches long and generates a muzzle velocity of 830 meters/2723 feet per second. The cyclic rate is between 700 and 800 rounds per minute, and the export potential of the model has been raised by the availability of a vz59N variant chambered for the NATO rather than Soviet 7.62 mm round.

The Soviet army uses separate general-purpose and squad support machine-guns, the former being the PK chambered for the oddly antiquated 7.62 mm x 54R rimmed round, and the latter the RPK chambered for the 7.62 mm x 39 round as used in the AK series assault rifle to which the RPK is closely related. Given the lavish scale on which Soviet small arms and ammunition are produced and distributed, the inherent logistical supply problem is clearly not an insurmountable difficulty.

The PK is in fact one of a series of weapons, a highly capable and still much-feared weapon that first appeared in 1946 with a fluted heavy barrel for the medium machine-gun role. The gun itself weighs 9 kg/19.8 lb, and is usually fitted on a 7.5 kg/16.5 lb tripod on which it is fed by 100-, 200- or 250-round belts for a cyclic rate of about 700 rounds per minute at a muzzle velocity of 825 meters/2707 feet per second with a 658 mm/25.9-inch barrel. In sustained fire frequent barrel changes are advised, but like many other Soviet automatic weapons the PK has a chrome-plated bore to reduce wear. The design is simple yet effective, as are all in-service Soviet weapons, which are designed to be

The trio of small arms that served the US Army so well in World War II was the M1 rifle (left), the M1928 (and M1 successor) sub-machinegun (center) and M1911A1 self-loading pistol (right), all seen in a 1945 propaganda photograph.

Above:
The provenance of the Heckler and Koch G3 can be clearly seen in this illustration of a Spanish infantryman with his 7.62 mm CETME Model 58 assault rifle.

Far right:
A measure of the M2 heavy machine-gun's capability is its continuing use as part of the secondary armament of such advanced weapons as the M1 Abrams, the most recent main battle tank in US service.

"soldier-proof". Other models in the series are the IKM product-improved and lighter variant, the PKS tripod-mounted AA variant, the PKM bipod-mounted squad support weapon, the PKMS tripod-mounted AA variant of the PKM, and the PKB with the standard butt/trigger group replaced by spade grips and a butterfly trigger. The Type 80 is the PK produced by China.

The RPK is solely a squad support weapon, and first appeared in 1966 as a derivative of the AKM assault rifle. The Chinese copy is designated the Type 74. Apart from a longer and more substantial barrel plus a light bipod, the RPK is identical with the AKM. Ammunition feed is from a special 75-round drum, or 30- or 40-round box magazines can be used. The cyclic rate is 660 rounds per minute, but a more practical rate, heavily impressed on trainees, is 80 rounds per minute as the barrel cannot be changed.

The USSR is now changing over to a new standard 5.45 mm/0.215 inch x 18 round, and the RPK has been adapted for this cartridge as the RPK-74. This is identical with the RPK in all respects other than barrel bore etc.

This tendency towards smaller rounds for assault rifles is also readily discernible in the West, and squad support weapons have thus been evolved to work in concert with these individual weapons. One of the most capable is another FN product, the Minimi chambered for the new NATO SS109 5.56 mm round. This is designed for a greatest effective range of 400 m/440 yards, now considered the maximum for battlefield engagements. The Minimi is a neat weapon with belted ammunition (100 or 200 rounds) in a box under the gun body, though the type can also use the 30-round box magazine of the US M16A1 assault rifle fitted into the receiver under the belt-feed guides after the belt has been removed. The box magazine is fitted with a simple indicator of the number of rounds remaining.

Many of the Minimi's features are adaptations from the MAG (typical being the gas regulator and the quick-change barrel system), but the gas-operated bolt system has rotary locking, and moves inside the receiver on twin rails to provide absolutely smooth and vibration-free movement, a factor which has helped to make the Minimi accurate and reliable. Overall length is 1050 mm/41.3 inches, and the weapon complete with bipod weighs 6.5 kg/14.3 lb, rising to 9.7 kg/21.4 lb with 200 rounds of ammunition. Muzzle velocity and cyclic rate are 915 meters/3002 feet per second and between 750 and 1,000 rounds per minute respectively. There is little doubt that the Minimi is an excellent weapon and the standard against which all comparable weapons must be judged. The type has been adopted by the US Army as the M249 Squad Automatic Weapon, and other large sales must inevitably follow.

A competitor in third-world and Asian markets must be the Ultimax 100 produced by Chartered Industries of Singapore, and chambered for the US M193 5.56 mm round, which fires a slightly lighter and shorter round than the SS109. This weapon is in effect the machine-gun version of the M16 series assault rifle (and thus a parallel to the RPK), but is notably light (6.5 kg/14.33 lb complete with 100-round ammunition drum under the body) and with a high muzzle velocity of 990 meters/3248 feet per second. The 100-round drum can be replaced by 20- or 30-round box magazines. The Ultimax 100 is available in a Mk 1 form with a fixed barrel and a Mk 2 form with quick change barrel, and sales have already been made to the Singapore armed forces. Several other armies are trying the type, and sales will inevitably follow.

The British contender in this market is the gas-operated L86 Light Support Weapon, the squad support counterpart of the L85 Endeavour assault rifle. Original plans were for these two complementary weapons to use the British-developed 4.85 mm/0.19-inch round, but the attractions of commonality prevailed and the two members of the Enfield Weapon System (otherwise Small Arms 80, or SA80) are now chambered for the 5.56 mm SS109 NATO round held in the 30-round magazine of the M16 assault rifle. The main differences between the L86 and L85 are the former's bipod, heavier barrel and pistol grip under the "stock", though the SA80 series is of the new "bull-pup" design concept based on a straight line between muzzle and butt, and with the trigger group forward of the box magazine.

Bull-pup weapons

The weapons discussed above are generally conventional in layout, even when revamped to use the new small-caliber rounds currently in favor. However, there are a number of considerably more advanced designs on the market and entering widespread service, known generically as "bull-pup" designs. These are characterized by a straight design taking the recoil force through to the firer's shoulder without any angles, so reducing the tendency of the muzzle to rise, and by the positioning of the trigger group in front of the magazine, decreasing length and thus weight. Weight is also lessened by the liberal use of nylonite or comparable plastic materials and light alloys in less-stressed components. Thus a typical "bull-pup" design may have only its barrel and receiver made of high-strength steel, which reduces both weight and manufacturing cost and time. It is no exaggeration to say that the new generation of assault weapons is transforming the military world as much as the machine-gun or bolt-action rifle did. The tendency is confirmed by the availability of whole families of weapons based on the same core design, so that armies can field the gamut of infantry small arms with a much reduced spares holding and its associated logistical infrastructure.

The three leaders in this new field are the Austrian Steyr AUG (Armée Universal Gewehr), the British L85

One of the best modern weapon combinations is that of the M16A1 assault rifle with the M203 40 mm grenade-launcher tucked under its forebody with a separate trigger just in front of the M16's magazine. A useful range of 40 mm grenade types enables the infantryman to engage his target in a variety of ways.

Top left:
An extraordinary fitting developed for the StG44 was the Krummlauf muzzle attachment, which allowed the weapon to be fired around corners with the aid of a mirror sight. The device worked, but was of no significant value to Germany by the end of World War II.

Bottom left:
The L7A1 is the main production variant of the FN MAG machine-gun used by the British army. The type is shortly to be relegated to the sustained-fire role as the L86 squad support variant of the L85 assault rifle begins to enter service.

Right:
The difference in size between the L1 (left) and L85 (right) can clearly be seen in this comparative view. The whole range of British army drill is being revised to accommodate the new weapon, which may lack the solid appeal of its larger predecessors, but has greatly improved firepower and ammunition capacity.

Endeavour, and the French FA MAS, all chambered for 5.56 mm rounds. The AUG has the look of total modernity, even of science fiction, and has such advantages as a clear plastic magazine (holding 30 rounds) so that the firer can see at a glance the number of rounds left, and modular construction of units such as the trigger group so that repairs are simple and quick. The weapon can be stripped with great simplicity, and the chromed barrel interior makes this essential component long-lasting and easy to clean. The standard x1 optical sight (which, like that of the M16 series, forms a carrying handle) can be easily replaced by a sniper's sight, night sight etc, and the pointability of the weapon is excellent as the right hand holds a shaped pistol grip under the trigger and the left another handle slightly farther forward. The AUG is 790 mm/31.1 inches long and weighs 4.1 kg/9 lb with 30 rounds, and the other basic figures are a cyclic rate of 650 rounds per minute and a barrel length of 508 mm/20 inches. Sales have been made in the Middle East and South America as well as to the Austrian army, and the prospects of this far-sighted weapon are very good, especially as other versions such as a light machine-gun are available.

The FA MAS is also striking in appearance, even if totally different from the AUG in looking stocky rather than sleek. Again a distinctive feature is the sight-cum-carrying handle of massive appearance, but there is no forward handle, the firer having a wide but well-shaped grip under the barrel for his left hand, plus a conventional pistol grip for the trigger hand. The butt is angular but very sensibly designed, and a light bipod is standard. Simplicity of manufacture and maintenance were stressed at the design stage, so the FA MAS is somewhat crudely finished (plastic being used wherever possible) and lacks even a chromed bore. The type is now well established in French

Left:
Conditions in Vietnam were often singularly taxing for men and weapons: a member of a US Navy SEAL (SEa-Air-Land) special unit makes his way through waist-deep mud with his Stoner 63 light machine-gun, complete with boxed ammunition belt.

Below:
A British paratrooper demonstrates how the collapsible sub-machinegun (in this case a Sten) can be tucked safely inside the man's parachute harness.

service, and has proved popular and reliable. Overall dimensions are small (length a mere 757 mm/29.8 inches), and weight is low at 4 kg/8.8 lb with a loaded 25-round box magazine. Rate of fire and muzzle velocity are 900 to 1000 rounds per minute and 960 meters/3150 feet per second respectively. The delayed blow-back operation is simple and so reliable, and the firer has the option of single-shot, automatic and three-round burst fire.

The L85 Endeavour, otherwise known as the Individual Weapon of the SA80 family, has been under development for some years and started to enter service at the end of 1985. The weapon looks distinctly aggressive. The SUSAT optical sight (replaceable by other sights for sniping and night firing) is over the body of the weapon, while the firer has a comfortable foregrip and the usual pistol grip under the trigger. As with other "bull-pup" designs, the 30-round magazine is well to the rear of the weapon, snuggling against the firer's right forearm when the weapon is shouldered. As with other "bull-pup" designs, the standard weapon can be used only by right-handers: the ejection port is level with the firer's cheek, meaning that a left-handed firer would be burned here. Overall length is 30.3 inches/770 mm and barrel length 20.4 inches/518 mm to give a muzzle velocity of 900 meters/2965 feet per second and a cyclic rate of 700 to 850 rounds per minute. The loaded weight is 10.14 lb/4.6 kg, and the whole feel of the

Above:
The superb Browning M2 heavy machine-gun is still in production and widespread service 65 years after its introduction, though mostly in vehicle installations such as the US M578 light armored recovery vehicle.

Right:
The M60 machine-gun in action aboard a US river patrol craft during the Vietnam War.

weapon is excellent with minimum recoil and good pointability for maximum accuracy. The receiver and stock are metal pressings, and the foregrip is made of nylon.

The Soviets may well follow this latest Western lead, but by then there is a possibility that the next step may be at hand, its forerunner being the Heckler and Koch G11, a truly remarkable prototype. The core of the new concept embodied in the G11 is caseless ammunition, which means that there is no spent cartridge case to be extracted and ejected. The 4.7 mm/0.185-inch type developed by Heckler and Koch uses a rectangular block of propellant tipped by the bullet, and is apparently sufficiently flammable for ease of firing but not so flammable that the round will "cook off" if left in a hot breech. The G11 was designed around this ammunition, and the gun is enclosed in a long outer casing on which the only projections are the combined carrying handle and sight above, and the pistol grip and trigger below. Two detachable magazines are carried above the barrel, each holding 50 rounds, which are fed into the chamber by a rotary device that keeps out dirt: once a magazine is exhausted it is removed, discarded and replaced by a fresh magazine. There is no conventional ejection port, but the loading disc can be manipulated to open a port through which any misfired round can be extracted. A selector switch by the trigger can be set for single-shot, three-round burst and automatic fire, the burst rate being 2000 rounds per minute and the automatic rate 600 rounds per minute. The muzzle velocity is estimated at 930 meters/3050 feet per second, and other details are an overall length of 750 mm/29.5 inches, a barrel length of 540 mm/21.3 inches, and a loaded weight of 4.5 kg/9.9 lb. The G11 is an experimental weapon (few details of the propellant or of the feed and operating systems have been revealed), but Heckler and Koch appear to have solved the basic problems of caseless ammunition, and it is the surest indication yet of the way ahead.

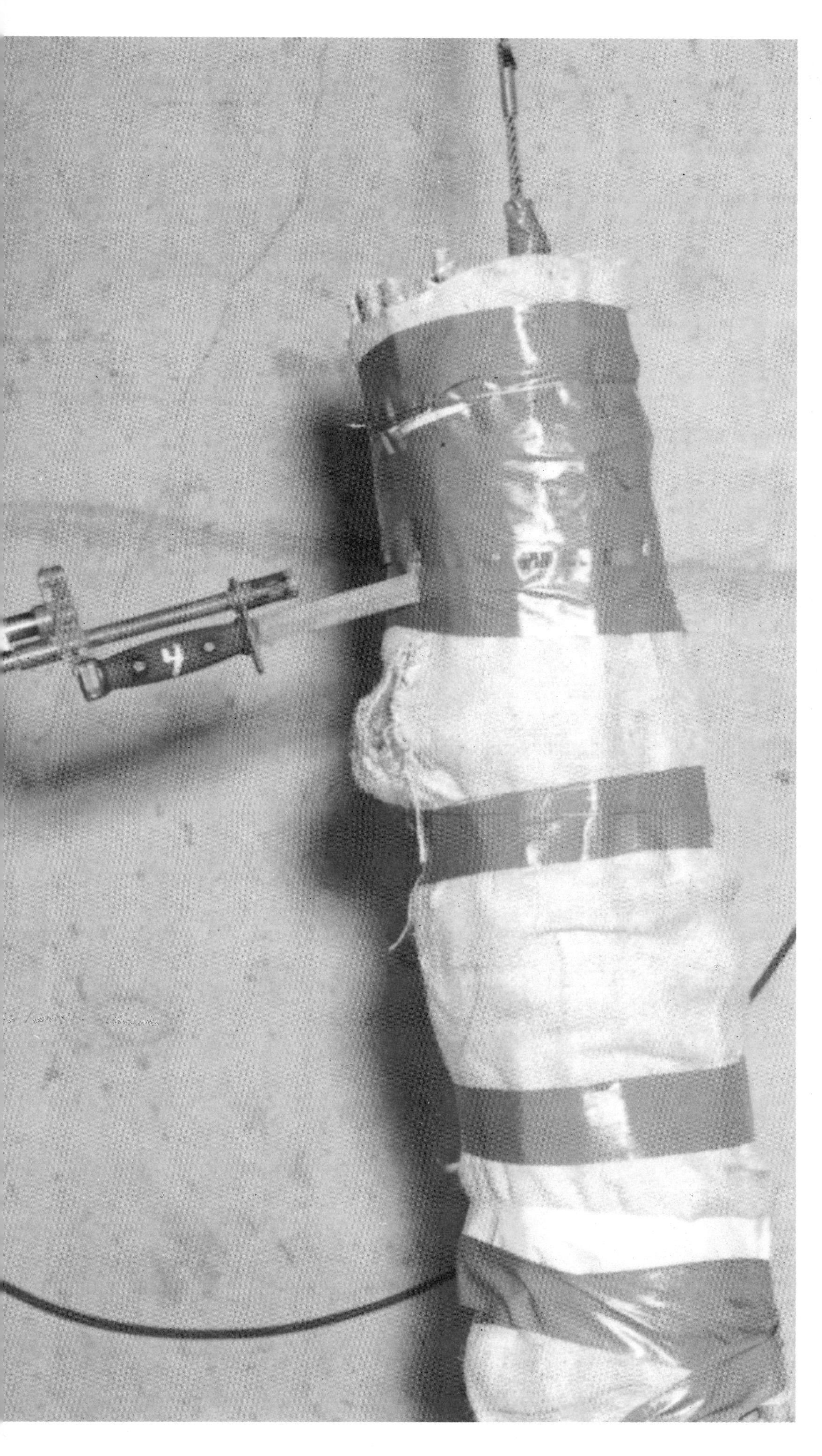

A member of the Stoner rifle series, the 5.56 mm XM22E1 was evaluated by the US services but not standardized for full service.

Sub-machine-guns, pistols and other weapons

Multi-shot capability and high volumes of fire have long been the pre-occupation of small arms designers and operators alike, and the two factors are closely related. As noted previously, such weapons had to wait for the development of the right ammunition in the form of metallic cartridges to make feasible the practical development of the revolver pistol, and of the right propellant in the form of "smokeless" nitrocellulose for the practical development of the self-loading and automatic weapon. In both instances, "practical" is a significant word, for some earlier schemes in both fields had worked to limited degrees. But once nitrocellulose-filled metal-cased cartridges were available, development of the self-loading and then the fully automatic weapon proceeded rapidly to produce the semi-automatic pistols and automatic machine-guns discussed above.

As with other types of small arms, such weapons were heavily influenced by trench warfare in World War I. Pistols, for example, had been regarded as suitable only for officers and second-line troops, but trench fighting revealed all too dramatically that bolt-action rifles were too long (and even longer when fitted with a bayonet) for such fighting, and too slow to reload. In these circumstances the revolver pistol was an ideal weapon for it was handy and could deliver six large-caliber bullets in quick succession; earlier misgivings about the pistol's accuracy were proved largely irrelevant as ranges were so short during trench fighting that the firer could hardly miss. Semi-automatic pistols had even greater magazine capacity and rate of fire, but were generally regarded as unreliable in extreme situations.

Below:
The Beretta 92F in 9 mm Parabellum caliber, one of the best self-loading pistols currently in production, forms the design basis for the pistol selected to replace the venerable 0.45-inch Colt M1911A1 in service with the US forces.

Left:
Weapons such as the MP38/40 sub-machinegun, were designed for mass production, and were very important to the German war effort as well as providing second-line troops with useful firepower.

Four members of the Webley and Scott pistol family. Clockwise from top left: the Mk I introduced during 1887 in 0.455-inch caliber, the Mk VI adopted during 1915 in 0.455-inch caliber, the Mk VI training model in 0.22-inch caliber, and the Mk IV accepted for service during 1899 in 0.455-inch caliber.

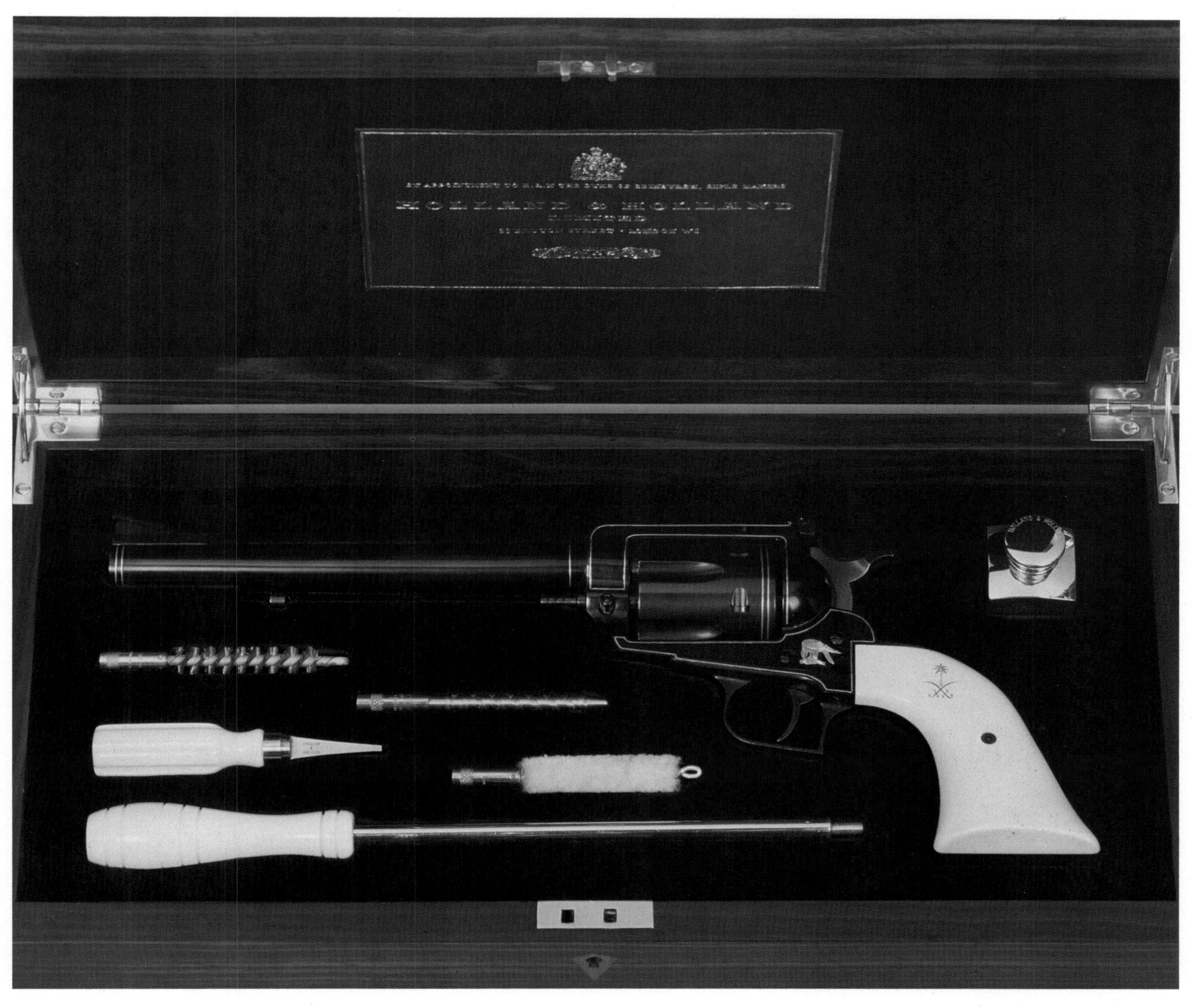

Above:
Perhaps less ostentatious than the Smith & Wesson revolver, this Holland and Holland offering is the more impressive for its elegance.

Left:
Designed in the classic revolver mold, the stainless steel Ruger Redhawk is constructed to the highest standards, and chambered for the exceptionally powerful 0.44-inch Magnum round.

Revolvers

Thus the design and manufacture of revolver and semi-automatic pistols came in for close examination between the world wars, and the result was a batch of reliable and, under the right circumstances, useful pistols. Great progress was made with the reliability of semi-automatics, and the revolver began to slip into a secondary position, at least as far as front-line troops were concerned. The revolver was still widely used for second-line duties, however, by units such as the military police who preferred the massive stopping power of revolver bullets to the semi-automatic's higher rate of fire and overall compactness.

Typical of such revolvers was the Enfield No.2 Mk1, a six-shot 0.38-inch/9.65 mm weapon weighing 1.7 lb/0.77 kg, firing a heavy bullet at 600 feet/183 meters per second from a 5-inch/127 mm barrel. The weapon in its definitive No. 2 Mk 1* form was a double-action weapon in which pulling the trigger cocked and then released the hammer to fire the weapon without the clothes-catching protrusions of the original single-action No.2 Mk1 in which the hammer had to be cocked manually before a pull on the trigger released it. The Enfield revolvers were heavy hinged-frame weapons which automatically ejected the cases when the gun was broken, but were utterly reliable man-stoppers at short ranges. Comparable weapons were the Smith & Wesson 0.38/200, the Smith & Wesson M1917, the Colt M1917 and the French Model 1892. Since the end of World War II, however, the revolver has faded quite rapidly from the military scene, and is now used only by military police units.

In the civil and paramilitary market the revolver survives as a sporting, personal defence and police weapon. The country most involved in such design and manufacture is the USA where new and highly capable weapons, and also copies of celebrated older models, are made. Typical of these weapons are the Charter Arms Undercover Model (a short-barrel 0.38 Special five-shot weapon available in single- or double-action models), the Colt Trooper (a medium-barrel 0.38 Special or 0.357 Magnum six-shot weapon available in single- or double-action models), the Smith & Wesson Bodyguard (a short-barrel 0.38 Special five-round weapon in single- or double-action models) and the Ruger Blackhawk (a long-barrel 0.357, 0.41 or 0.457 Magnum six-shot single-action weapon). There are many other such models, most of them optimized for police work with short barrels and shrouded hammers, and it seems likely that such revolvers will perform usefully in these roles for years to come.

Semi-automatic pistols

World War II proved the semi-automatic pistol in combat and generated massive orders for it. The best of the older generation remained in service (for example, the Pistole 08 and the Colt M1911A1), while newer and generally more handy weapons were the 7.62 mm Tokarev TT-33 from the USSR, the 9 mm Walther P38 and PP-PPK series from Germany, the quite superb 9 mm Browning Hi-Power from FN in Belgium and the 9 mm Beretta Model 1934 from Italy.

Left:
The world's first successful sub-machinegun was the Bergmann MP18, seen here in the hands of a German paramilitary policeman after World War I. Note the two "snail" type magazines in front of the man's left arm.

Bottom:
The Federal Riot Gun is typical of current design practice in this field. The weapon has a caliber of 37 mm, a weight of 7.6 lb/3.4 kg and is 29 inches/737 mm long. The weapon is of the double-action type with no external hammer to catch on clothing, and can fire the full range of anti-riot munitions to a range of 100 yards/91 m.

Below:
The Power Staf is a new departure in riot-control weapons; it is designed to prod rioters with an extremely rapid to-and-fro motion of the weapon's tip through a gas-operated mechanism, causing severe bruising.

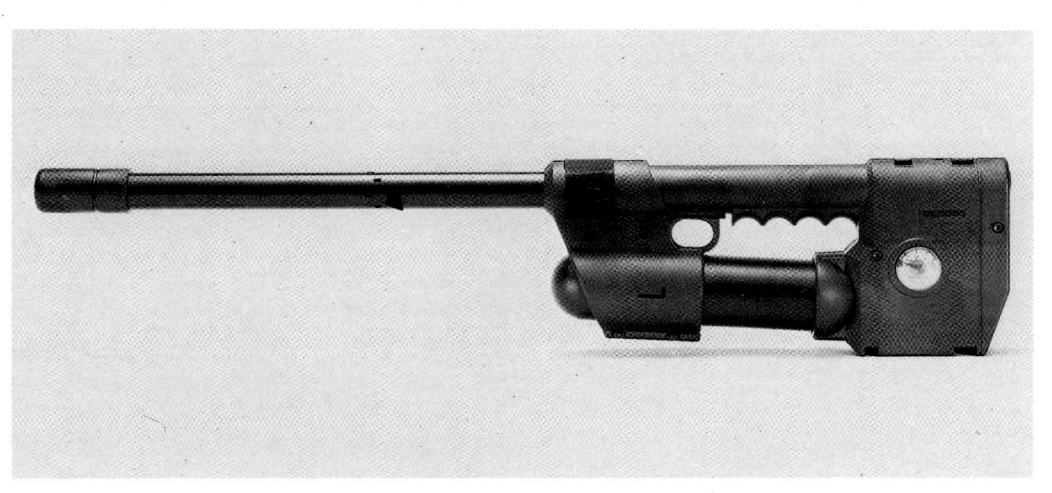

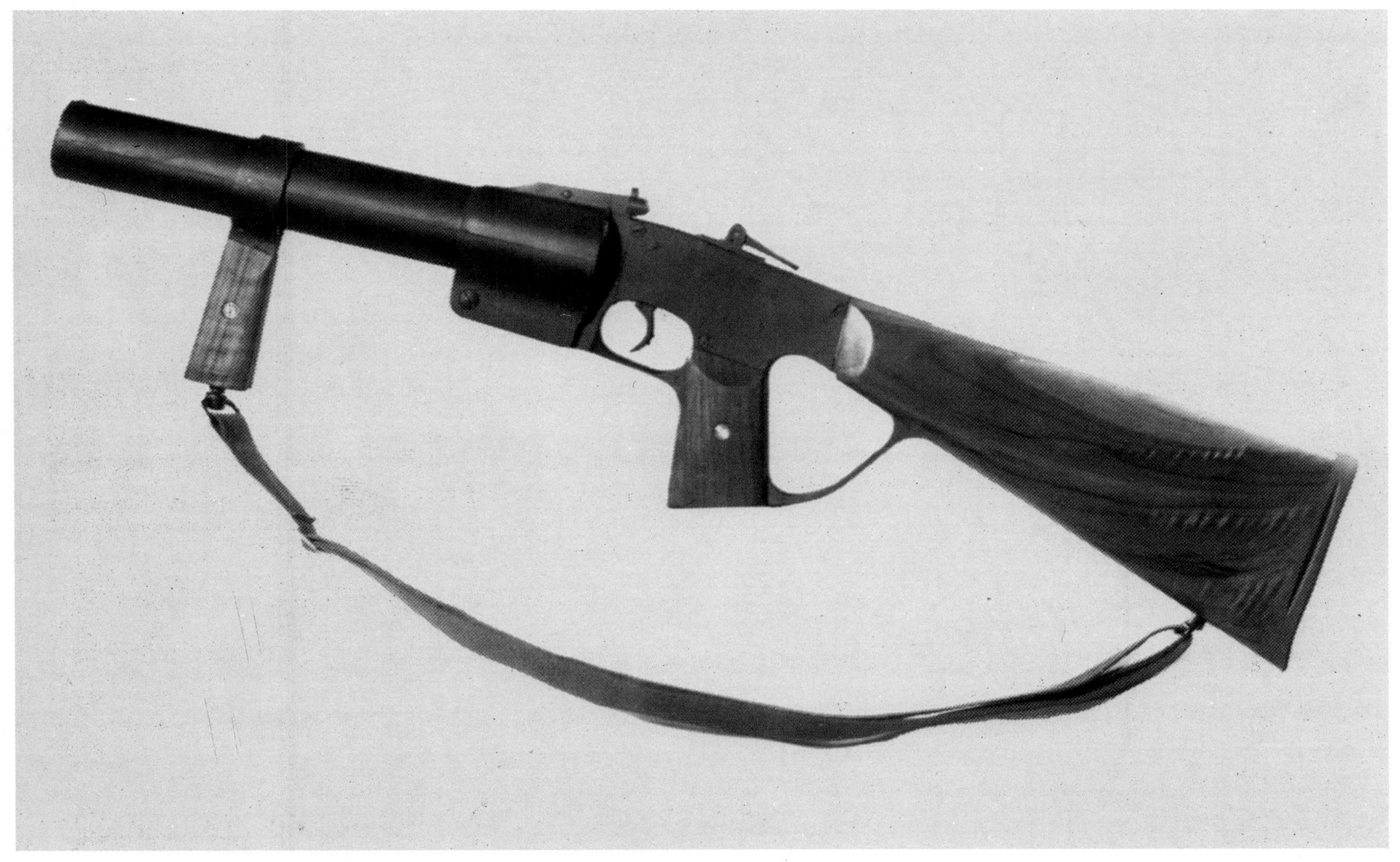

The Centrum Free Pistol from East Germany is a good example of the 0.22-inch precision target-shooting pistol type, with the butt carefully designed for weapon balance and maximum stability of the shooting hand.

9
0 9 8 7

The P38 was designed to succeed the legendary P08, but failed to replace its predecessor, though it was an excellent weapon in its own right with an eight-round magazine and 124 mm/4.9-inch barrel firing a Parabellum round at 350 m/1150 ft per second. The PP and PPK were designed as medium- and short-barrel police pistols, but found a ready market with the military and the paramilitary forces of Nazi Germany. The genuine worth of the type, which was introduced in 1929, is reflected in its being still in production and worldwide service. The PPK was available in a number of calibers, and fired a moderate bullet from an 86 mm/3.4-inch barrel at a velocity of 280 m/920 ft per second. The magazine held seven rounds. Altogether more powerful, but also heavier and with far greater recoil, the Browning HP had a 13-round magazine in the butt, and fired 9 mm Parabellum rounds from a 112 mm/4.4-inch barrel at 354 m/1160 ft per second. Semi-automatic pistols continue to be developed, many of them variations on an older theme,

The world's first sub-machinegun was the twin-barrel Villar Perosa with the underpowered 9 mm Glisenti cartridge, produced in Italy from 1915. The weapon was

too large for a sub-machinegun, and was often used as an infantry support weapon (almost a light machine-gun) with a shield to protect the crew.

and the most important of these is the new Beretta that has been selected to replace the faithful M1911A1 in US service.

Sub-machineguns

Trench fighting in World War I also demanded greater volume of fire than could be delivered by pre-war weapon types, and here the sub-machinegun entered the scene as a hybrid type designed to combine the handiness of the semi-automatic pistol with the automatic fire of the machine-gun. It was, of course, impossible to develop such a weapon with the full-power rifle cartridge, for the recoil would have made it totally unmanageable, so the weapon that emerged was designed for a pistol-power round held in a large-capacity magazine, with a recoil-operated action to provide full-automatic fire. Such a weapon was clearly short-ranged and intrinsically inaccurate, but this was of little consequence in trench fighting where the firer wanted on

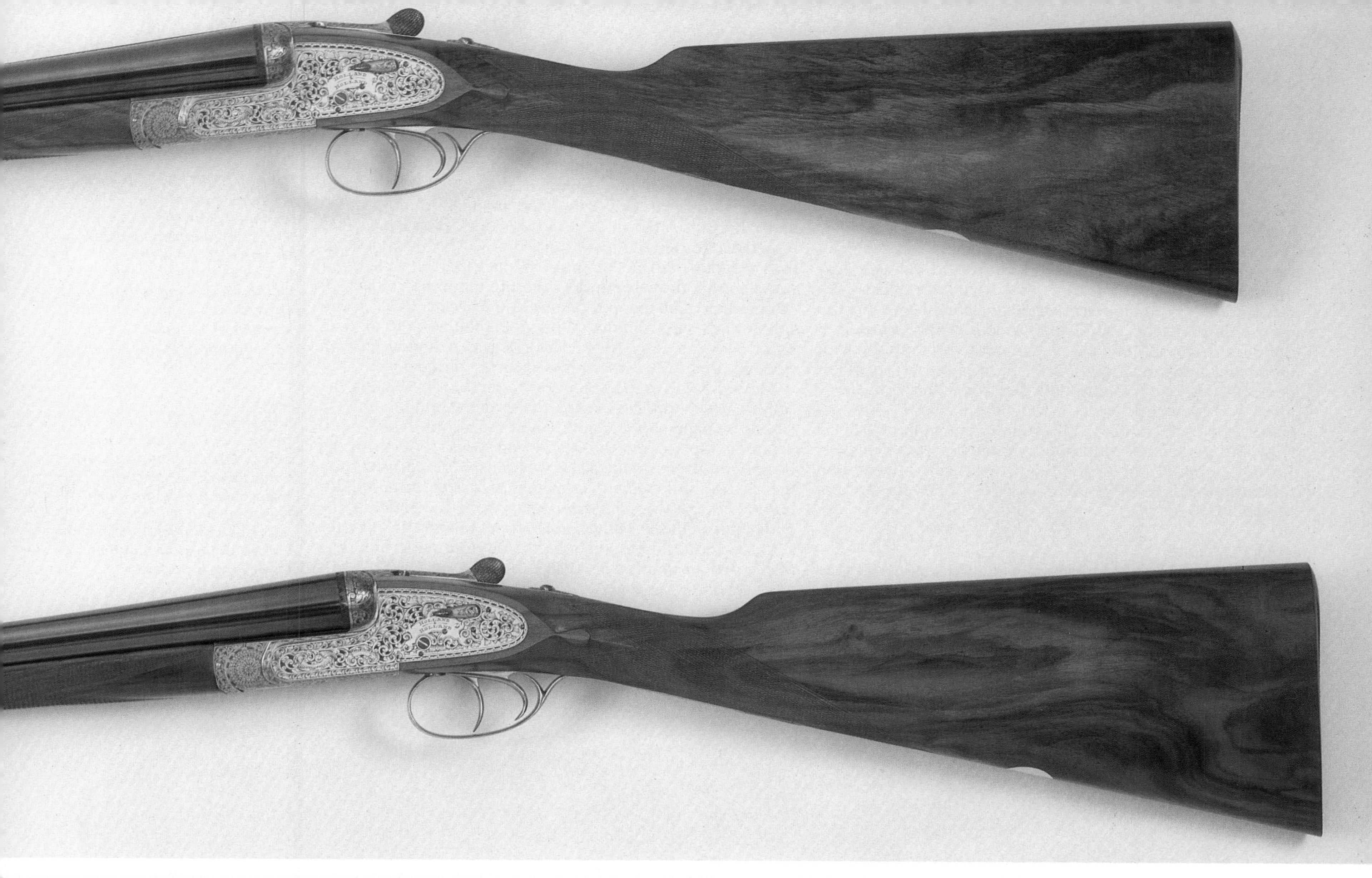

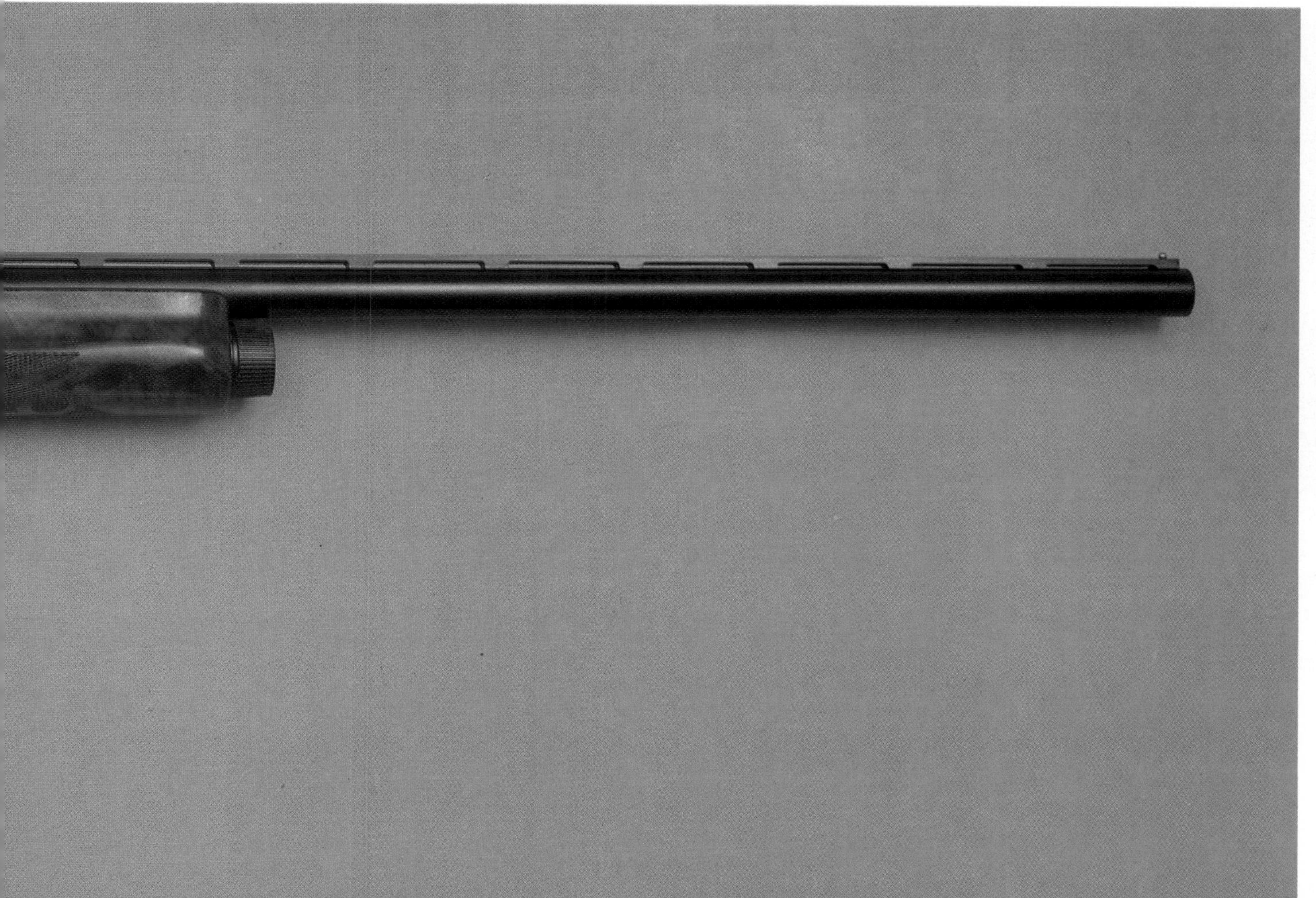

Above:
Typical of the very best modern shotguns are the side-by-side twin-barrel weapons produced in the UK by Holland and Holland. Noted for their superb finish, excellent balance and exquisite craftsmanship in metalwork and wooden furniture, they are thoroughly practical as sporting weapons.

Left:
The Remington 1100, typical of modern self-loading shotguns. Gas is tapped off the barrel to operate a mechanism that ejects the spent case and loads a fresh round from the four-round tubular magazine.

demand to generate a mass of fire at very close range.

ITALY

The first such weapon was an Italian product, the Villar Perosa which appeared in 1915. The weapon was designed for use in aircraft, but the Italian authorities far-sightedly saw that its best application was in infantry use. The weapon was a double-barreled device chambered for the poor 9 mm Glisenti round (the same cartridge as used in the Beretta Model 1934 pistol, and not to be confused with the higher-powered 9 mm Parabellum round). Capable of operating only on full-automatic fire, the Villar Perosa was a delayed blowback weapon weighing 6.5 kg/14.3 lb and firing at a cycle rate of 1200 rounds per minute from a 25-round detachable box magazine above each receiver. From this initial model Italy developed two other 9 mm Glisenti models in World War I (the OVP and the Beretta Model 1918).

ANS POWERS

Design leadership had already passed to Germany with the Bergmann-designed Maschinenpistole 18, in effect the world's first true sub-machinegun. This was a simple blow-back-operated weapon chambered for the 9 mm Parabellum round, at first contained in a 32-round "snail" magazine but later in 20- and 32-round box magazines inserted in the left of the weapon's body. The 815 mm/32.1-inch weapon had a 200 mm/7.87-inch barrel which was sufficient for the weapon to generate a muzzle velocity of 365 meters/1,200 feet per second and a cyclic rate of 350 to 450 rounds per minute. With a loaded weight of 5.25 kg/11.6 lb the fully-stocked MP18 was a handy enough weapon, and fully foreshadowed future designs.

The type was put back into production in 1928 as the MP28, which differed from the original model only in having a single-shot capability and in various detail modifications. Next came the MP34 and MP35, still from the Bergmann design team. The weapons were derived from the MP18, but refined considerably. The primary differences were the magazine housing in the right of the body, and a double-pull trigger system, the first position for a single shot and the second for full-automatic fire. The MP35 was an improved MP34.

But in 1938 there arrived the classic MP38, which was

Right:
The Colt M1911A1 0.45-inch self-loading pistol was developed from the basic M1911 and adopted in 1926. An extremely powerful weapon well-suited to the military, it is about to be replaced only because current examples of this out-of-production weapon are wearing out.

Below:
Competition models of standard self-loading pistols, such as this Belgian-produced 9 mm Browning, receive special attention in the final machining and assembly to ensure that all tolerances are as close as possible and that the weapon is perfectly balanced.

CARTRIDGES CAL .45
WIND TO 9 CLICKS

The classic Thompson (top) in M1928A1 form. The weapon is fitted with the 20-round box magazine, and lower left is the alternative 50-round drum magazine. Less successful was the Reising Model 50 (below), a weapon developed in the late 1930s in 0.45-inch caliber.

Left:
Genuine target-shooting pistols such as the 0.22-inch Browning International Mk 2 are fitted with special grips to ensure a comfortable grasp, and are balanced to help the firer maintain a steady aim.

Above:
The standard self-loading pistol used by the Soviet forces in World War II was the Tokarev TT-33. This weapon gave an officer a measure of firepower but left him free to give hand signals to his unit.

more representative of the sub-machinegun as it is known today: the Bergmann and other weapons of its type had been made with the same care as contemporary rifles, but the MP38 was designed for ease of mass production. The MP38 used a folding wire-framed butt, the body was produced from simply-made sheet metal stampings, and the bolt was machined only where absolutely needed. The result was a thoroughly utilitarian and threatening weapon that worked extremely well in its designed role. A pistol grip at the rear of the receiver and the housing under the forward part of the body for the vertical 32-round magazine provided two good handholds, and the weapon could thus be pointed almost instinctively. Further refinement produced the MP38/40 and the classic MP40 with yet further simplified manufacture. This last weighed 4.7 kg/10.4 lb loaded, was 833 mm/32.8 inches long with the stock extended, and could fire 9 mm Parabellum rounds at 500 rounds per minute, and a muzzle velocity of 365 meters/ 1200 feet per second from the 251 mm/9.9-inch barrel. Further German work on sub-machineguns was unnecessary during World War II as the MP40 was fully adequate to its task and as the assault rifle arrived in the form of the StG44 to replace the rifle and sub-machinegun for close combat.

The Italian sub-machineguns of World War II were two beautifully made and fully-stocked weapons of the older design school, the Beretta Model 1938 and Model 1938/42. These weapons originally used a Beretta-developed 9 mm round, but were later revised to take the standard 9 mm Parabellum round. The Japanese also used a stocked weapon, the 8 mm Nambu Type 100 which was an adequate but hardly outstanding weapon with too complex a feed mechanism and a highly curved 30-round box magazine.

The Ingram type sub-machinegun series is extremely compact and easily made, the only machined component being the barrel. Above: the Model 10 for the 0.45-inch cartridge; below: the Model 11 for the 0.38-inch round. Both types have retractable wire stocks, exceptionally high rates of fire, and can be fitted with suppressors to reduce the noise signature.

SAFE
FIRE

UK

On the Allied side the British moved from the fully-stocked 9 mm Lanchester with its 50-round magazine to the completely utilitarian and magnificently cheap Sten gun, also in 9 mm Parabellum caliber and carrying a 32-round box magazine. The Sten was designed in 1940 by Major R.V. Shepherd and H.J. Turpin and produced at Enfield (hence the name, from the designers' initials and location) in vast numbers for British and resistance forces, and the basic design lives on to this day in the L2 Sterling submachinegun. The typical Sten Mk II weighed only 8.2 lb/3.7 kg loaded, and was 30 inches/762 mm long. The cyclic rate was 550 rounds per minute and a muzzle velocity of 1200 feet/365 meters per second was attained with a 7.75-inch/197 mm barrel. The Sten was generally reliable as it lacked many moving parts, and manufacturing standards were as low as possible in a crude but cleverly designed weapon intended almost as a throw-away gun for close-quarter combat. Early models had a wire-frame stock, but as the crisis for the UK passed in the later stages of the war refinements such as a wooden stock were introduced.

USA

The Americans started the war with the Thompson gun, perhaps the most famous sub-machinegun of all time and best known as the "Tommy gun". The weapon was the brainchild of General John Thompson in 1918, and was designed for the standard US 0.45-inch pistol round, 50 of

Below:
The Hawk MM-1 is a multi-shot weapon for riot-control munitions, the 12-round spring-loaded rotary magazine being designed to enable the firer to disperse rioters with the use of specialist 38 mm munitions at fairly long ranges.

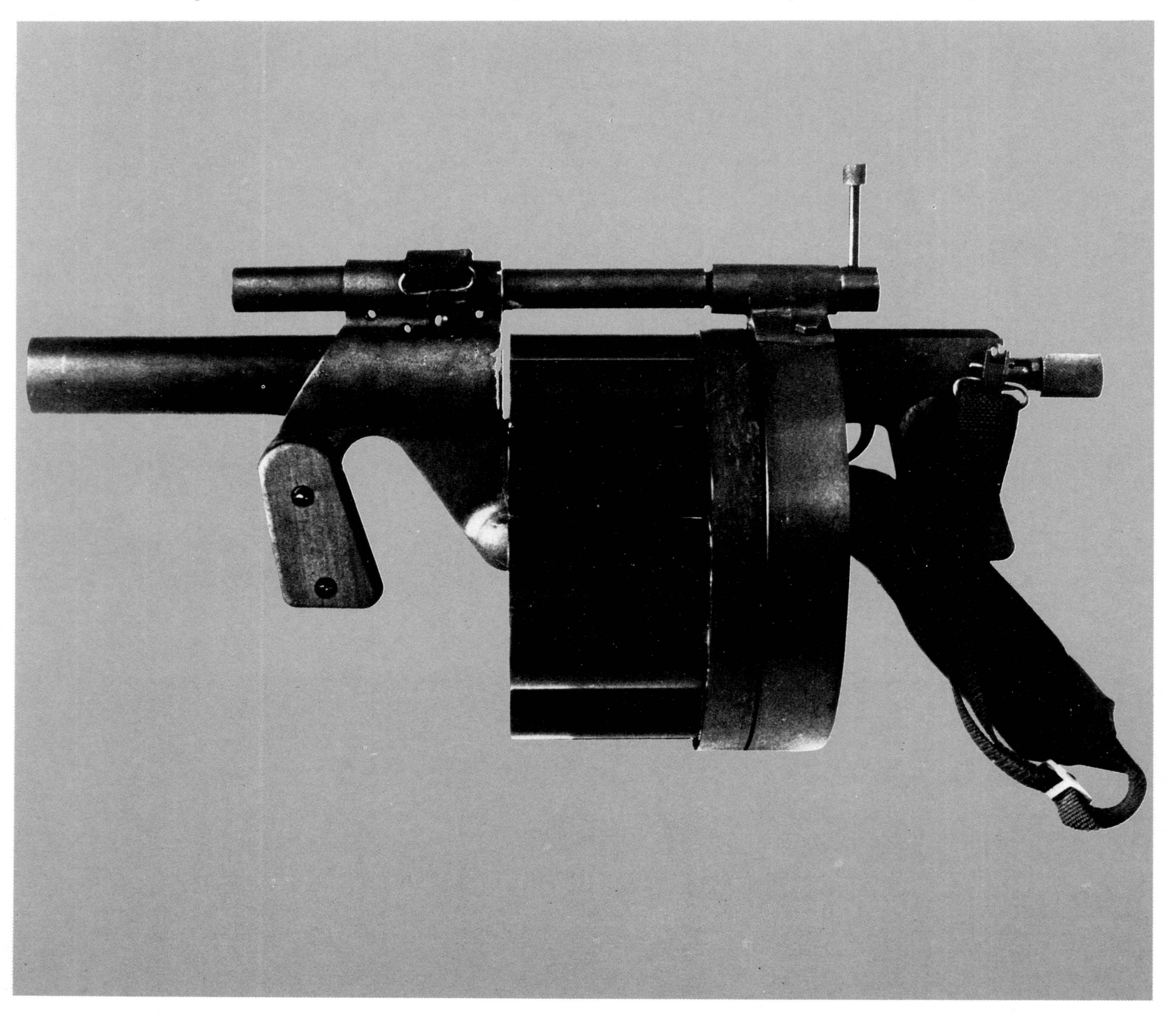

Left:
A weapon that proved popular with Australian troops was the 9 mm Owen sub-machinegun; it had the unusual features of a vertical overhead magazine and a plunger-operated quick-release mechanism for barrel changes.

which were held in a drum magazine under the forebody just in front of the pistol grip and trigger. Various trial models were produced in efforts to make the highly complicated blowback operation work properly before the first effective production variant appeared as the Thompson M1928. Few sales were made in the later 1920s and early 1930s, but in 1940 large-scale production was launched because of the crisis in Europe. Eventually the action was redesigned as a simple blowback type, and the troublesome drum magazine was replaced by a box magazine holding 20 or 30 rounds. This was placed in production as the M1 or M1A1 with some external modifications to ease manufacture. The weapon retained its wooden stock and foregrip, however, and thus weighed 10.45 lb/4.75 kg with the box magazine.

The M1 type was still expensive and fairly clumsy in combat, so in 1942 a far simpler weapon was standardized

Left:
The Czechoslovak sub-machinegun vz61, otherwise known as the Skorpion, is much favored by terrorist organizations as its compact shape makes for easy concealment. The type is chambered for a 7.65 mm round, and is only 271 mm/10.65 inches long, with the stock folded up and over the body of the weapon.

Above:
An officer of the US Army gets on with his task of bridge control without being hampered by his M3A1 sub-machinegun. Yet the weapon is available for emergencies, which is where it still scores. The passing infantryman is armed with the M21 sniping rifle.

Right:
The standard British sub-machinegun is the 9 mm Sterling L2A3, a linear descendant of the Sten used in World War II.

Clockwise from top left, are some outstanding self-loading pistols produced by the Germans in the period up to World War II. The Bergmann No.5 of 1897 (Bergmann's first military pistol) in 7.63 mm Bergmann caliber and sighted to a hopelessly unrealistic 700 m, the Luger P08 of 1908 in 9 mm Parabellum caliber, the Walther P38 of 1938 in 9 mm Parabellum caliber, and the Bergmann No.3 of the mid-1890s in 7.65 mm Bergmann caliber.

THE
"ROYAL"
MODEL

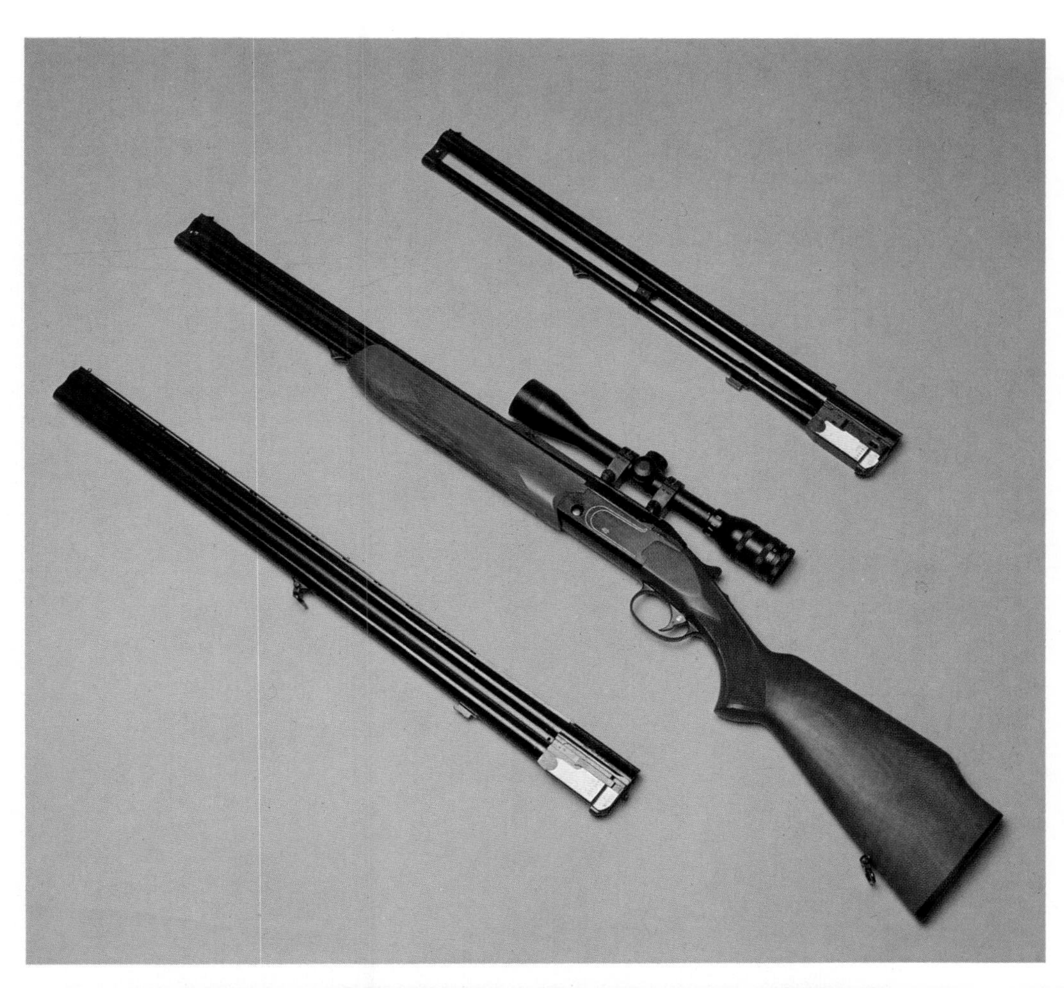

as the M3, though it was more generally known as the "grease gun". This may be regarded as the US equivalent of the Sten, and was designed like the British weapon for ease of manufacture rather than sophistication. The weapon had a retractable wire stock and was designed for the US 0.45-inch pistol round, but was also made in 9 mm Parabellum caliber for resistance forces. The initial model weighed 10.25 lb/4.65 kg when loaded with its 30-round box magazine, and had a cyclic rate of 350 to 450 rounds per minute. Operational experience and improved manufacturing processes led to the more capable M3A1 during the war, but the type did not enjoy great popularity and disappeared from service during the 1950s.

Far left, top:
Among the nearly perfect examples of the gunsmith's art must be sporting weapons such as the Holland and Holland Royal, with technical perfection allied to beautiful presentation.

Far left, bottom:
Apart from shotguns, Holland and Holland also make sporting rifles to the same high standards. This is a magazine rifle with detachable telescopic sight mount.

Left:
The Finnish Valmet 412 Shooting System is an advanced sporting weapon, based on a body to which can be attached any of four different stocks, seven shotgun barrels, nine combination gun barrels and six double rifle barrels. Other options are ejection or non-ejection, and single or double triggers.

Below:
Though not a top-level competition pistol, the Parker-Hale Star Model FR Target .22 Automatic is a good example of the target-shooting pistols used by gun clubs.

Soviet troops resting in a brief interval during the climactic Battle of Kursk, July 1943. Well to the fore are a trio of PPSh-41 sub-machineguns, each fitted with the 71-round drum magazine rather than the alternative 35-round box.

Top:
Though heavier than many other sub-machineguns, the Beretta weapons of this type are reputedly reliable, sturdy guns of excellent manufacturing standards. Top: the Model 1938A in 9 mm caliber; bottom: the lightened mass-production Model 1938/42 also in 9 mm caliber. Both have triggers, the forward one for single shots, the rear one for automatic fire.

Above:
The powerful cartridge that makes the M1911A1 so useful a man-stopper also produces a heavy recoil, and considerable training and experience is essential if the firer is to use the weapon to maximum effect.

Right:
The full-power rifle such as the British L7 series is too powerful for urban warfare while its weight and overall length make patrols in Northern Ireland more difficult than they would be with the new type of assault rifle.

USSR

The Soviets were great advocates of the sub-machinegun, though development of a suitable type had not begun until the early 1930s, resulting in the PPD-1934 soon refined into the PPD-1934/38. This weapon was derived in general from the MP18, and was chambered for the 7.62 mm Soviet pistol cartridge, of which 71 were held in a drum magazine derived directly from that of the Finnish Suomo m/1931 sub-machinegun. But the Soviets were quick to realize that they needed a mass-produced weapon equivalent to the German MP38 and British Sten, and so developed the classic PPSh-41, which had a loaded weight of 5.4 kg/11.9 lb and carried a 71-round drum or 35-round box magazine. The overall length was 828 mm/32.6 inches, and the cyclic

rate of 900 rounds per minute was achieved with a 265 mm/ 10.4-inch barrel that also generated a muzzle velocity of 488 meters/1600 feet per second. Though a stocked type, the PPSh-41 was very tough and ideally suited to Soviet tactics, though further simplification and the elimination of wooden features plus most of the machining produced the PPS-42. This was designed and made under the most adverse of circumstances inside beleaguered Leningrad during 1942, but was nevertheless a formidable weapon capable of a cyclic rate of 700 rounds per minute from a 35-round box magazine. The loaded weight was just 3.9 kg/8.6 lb.

It was clear by the end of World War II that the sub-machinegun was on its way out as a major-issue front-line weapon, for the assault rifle provided firepower and accuracy without significant weight or size penalties and with the added advantage of simpler logistics. Yet the sub-machinegun is still in widespread service, largely with specialist units, artillery and tank crews, and second-line forces that can make effective use of its special attributes (including silenced models). The French 9 mm MAT 49 picks up where the Sten left off in terms of reliability and simplicity of both manufacture and operation, the Czechoslovak 7.65 mm vz61 Skorpion is a minuscule weapon perhaps better described as a machine pistol, and the American 0.45-inch or 9 mm Ingram Model 10 is an extraordinarily compact design with the phenomenally high rate of fire of 1145 rounds per minute. But the two weapons that stand out head and shoulders above the rest as the current pinnacles of sub-machinegun design are the Israeli Uzi and West German Heckler and Koch MP5.

The 9 mm Uzi first appeared in the early 1950s, and draws heavily on previous designs for most of its features, though the genius of the designer has combined these into a formidably accurate and compact weapon only 650mm/ 25.6 inches long with a wooden stock (the weapon is considerably shorter with a folded wire stock) and weighing 4.1 kg/9 lb with a loaded 32-round box magazine. There is also a Mini-Uzi with the same basic capabilities as its larger brother, but only 360 mm/14.2 inches long with its wire stock folded and weighing 3.1 kg/6.8 lb with a loaded 20-round magazine.

Right:
One of the most effective, and most criticized, anti-riot munitions is the baton round, generally known as the "rubber bullet". This is a Schermuly 38 mm round of the type, though modern baton rounds are usually made of plastic.

Below:
The revolver was widely used by troops who could not carry a rifle, such as tank crews and headquarters staff. Here, a member of the British 1st Airborne Division uses his 0.38-inch Enfield Pistol No.2 Mk I to help fight off a German attack on the divisional HQ, during the Arnhem operation of September 1944.

Methven
KEILLOUR FOREST
PERTH
CRIEFF
Forgandenny
Dunning
AUCHTERARDER
Blackford
KINROSS
DOLLAR
CLACKMANNAN DISTRICT
BLAIRADAM FOREST
Kelty
Saline
ALLOA

The Tikka Model 55 heavy-barrel target rifle exemplifies its type, available in a number of calibers for competition shooting.

Top right:
The South African Stopper is a highly effective 37 mm riot-control weapon, neatly engineered to minimum weight and bulk and accurate to normal riot-control ranges.

Left:
The TRGG is a West German anti-riot weapon using the same principle as a flamethrower. It is designed for long-range projection of clouds of irritant agent over rioters. The weapon can deliver bursts or a continuous jet, and is meant to prevent the riot-control forces from coming into contact with the rioters.

Right:
Designed and produced in Italy, the MOD-T-22 is a long-range anti-riot system derived from the standard rocket-launcher concept, and intended for the bombardment of rioters with CS rounds at ranges beyond the rioters' capability to retaliate.

The MP5 is one of the enormous and highly diverse family of small arms produced by Heckler and Koch. The MP5 is typical of the series, a highly accurate weapon weighing only 3 kg/6.6 lb with a box magazine loaded with 30 9-mm Parabellum rounds. The overall length is 680 mm/26.8 inches, and the 225 mm/8.86-inch barrel produces a cyclic rate of 800 rounds per minute and a muzzle velocity of 330 meters/1083 feet per second.

It seems that the sub-machinegun will remain a viable weapon for special purposes over the foreseeable future, and further advances in design and materials will further lower weights and increase strength.

Riot control weapons

Another special-purpose small arms type is the riot-control weapon. Though these are generally simple single-shot weapons (using either a breaking frame or a pump action for rapid reloading), their use means that the designer has to take special care: designed for urban and paramilitary operations, the riot-control gun is usually a shotgun type firing birdshot or other light loads to disperse crowds or a modified flare gun/grenade-launcher type firing baton rounds or special riot-control munitions with a payload of tear gas or more modern equivalents such as CS. The former type is usually in 12-gauge pump-action repeater form so that the firer has at least five rounds at his disposal, and the latter usually has a caliber of 37 or 40 mm. The designer must ensure that muzzle velocities are not lethally high, or ineffectively low. A balance must be struck between these two failings, and it is also better if the weapon can be made to look menacing.

The riot-control gun is an increasingly important part of many armed forces' inventories, as the growing prevalence of civil unrest as a political tool means that many such situations cannot be contained merely by the police, requiring the army to be summoned. The tactics of riot-control are difficult and the forces of authority must have the right weapons (in terms of total reliability and nicely-judged impact) used with the right strength to control any situation that might otherwise get out of control. The accompanying illustrations give a good impression of the riot-control weapons now in use. The most notable fact about many of these types is that they are purpose-designed for the task, whereas previous weapons were generally conversions or developments of civil weapons (shotguns) or armed forces' ancillary weapons (flare guns and the like).

Index

Illustrations are shown by figures in **bold type**

Picture Credits

John Batchelor 4-5 **P Beretta, Italy** 135, 158 **Bibliothek Nationale, Paris** 20-21, 24 bottom **Bodleian Library, Oxford** 10, 11 top, 23 bottom **Browning Sports, Abingdon** 172, 176 **Crown Copyright (MOD-Army), London** 150 bottom **FFV Ordnance, Sweden** 124-125 **Guns Review, Hebden Bridge** 162, 166-167, 170-171 bottom, 187, 194-195 **Gunshots Collection, London** 2-3 **Heeresgeschichtliches Museum, Vienna** 6-7 **FN Herstal, Belgium** 140 **Ian Hogg Collection** 165, 180, 193, 196-197 **Holland and Holland, England** 1, 163, 170-171 top, 186 **Imperial War Museum, London** 29, 80-81, 84-85, 88-89, 92-93, 95, 96-97, 98-99, 101, 105 top, 108-109, 153, 164, 168-169, 192 **Kunsthalle, Hamburg** 14-15 **Military Archive and Research Services, Lincolnshire** 9, 12, 16-17, 24 top, 25, 30 bottom, 32, 48-49, 59, 66-67, 75 bottom, 94, 100 top, 106-107, 111, 115, 129, 138-139, 146-147, 150 top, 154, 159, 181, 182 top **National Army Museum, London** 11 bottom, 13, 18 top, 19, 22, 22-23, 23 top left and right, 28, 30 top, 38-39, 45, 46-47, 50 top left and right, 51, 53, 55, 56-57, 58-59, 60-61, 64-65, 68-69, 70 left, 71 bottom, 72-73, 76-77, 102-103 **Novosti Press Agency, London** 105 bottom, 176-177, 188-189 **Peter Newark's Western Americana, Bath** 26-27, 33, 34-35, 36-37, 40-41, 50 bottom, 54, 70-71 top **Salamander Photo Library, London** 18 bottom, 42-43, 44, 52, 62-63, 74 top, 75 top, 78-79, 82-83, 86-87, 90-91, 122-123, 126-127, 142-143, 160-161, 174-175, 178-179, 182 bottom, 184-185, 190 top **SIG, Switzerland** 131 **Soldier Magazine** 134, 151, 183, 191 **Steyr-Daimler-Puch, Austria** 130 **US Army, Washington DC** 114, 152, 155, 190 bottom **US Marine Corps, Washington DC** 121 top, 128

Front cover: **Handford Photography**
Back cover: **Gunshots Collection, London**

Officers of the Women's Royal Naval Service undergo instruction on firing a revolver, in this instance the distinctly powerful 0.455-inch Webley service type.

Two service weapons, the L4A1 sniper rifle (top) and the Individual Weapon (bottom, seen here in the original 4.85 mm caliber) sandwich the experimental EM2 rifle that pioneered the British "bull-pup" design.